TIME-DEPENDENT QUANTUM MECHANICS OF TWO-LEVEL SYSTEMS

Other Related Titles from World Scientific

Frontier Problems in Quantum Mechanics
by Lee Chang and Molin Ge
ISBN: 978-981-3146-84-6

The Geometry of Quantum Potential: Entropic Information of
the Vacuum
by Davide Fiscaletti
ISBN: 978-981-3227-97-2

Solid State Quantum Information — An Advanced Textbook:
Quantum Aspect of Many-Body Systems
by Wonmin Son and Vlatko Vedral
ISBN: 978-1-84816-764-3

Introductory Quantum Physics and Relativity
2nd Edition
by Jacob Dunningham and Vlatko Vedral
ISBN: 978-981-3228-64-1
ISBN: 978-981-3230-04-0 (pbk)

TIME-DEPENDENT QUANTUM MECHANICS OF TWO-LEVEL SYSTEMS

James P Lavine
Georgetown University, USA

World Scientific

NEW JERSEY · LONDON · SINGAPORE · BEIJING · SHANGHAI · HONG KONG · TAIPEI · CHENNAI · TOKYO

Published by

World Scientific Publishing Co. Pte. Ltd.

5 Toh Tuck Link, Singapore 596224

USA office: 27 Warren Street, Suite 401-402, Hackensack, NJ 07601

UK office: 57 Shelton Street, Covent Garden, London WC2H 9HE

British Library Cataloguing-in-Publication Data
A catalogue record for this book is available from the British Library.

TIME-DEPENDENT QUANTUM MECHANICS OF TWO-LEVEL SYSTEMS

ISBN 978-981-3272-58-3

For any available supplementary material, please visit
https://www.worldscientific.com/worldscibooks/10.1142/11052#t=suppl

Desk Editor: Ng Kah Fee

Typeset by Stallion Press
Email: enquiries@stallionpress.com

Preface

Welcome potential readers! Over the years I repeatedly stumbled onto sets of data with time dependencies. In my case, such data usually came from the transitions between the energy levels of a defect in crystalline silicon. Often the assumption of two or three energy levels was sufficient to explain the data. This led me to search for an accessible treatment of the theory of few-level systems. I found partial pieces scattered across multiple sources.

This book is my attempt to bring together an accessible treatment. It is aimed at people starting to delve into two-level systems. You will be the judge of whether or not I have succeeded. The material builds on what is covered in the first undergraduate Quantum Mechanics course. The mathematics is on a level with that of an undergraduate Math Methods course. Quantum Mechanics provides the framework through the Time-Dependent Schrödinger Equation. An electron or atom interacts with an electromagnetic field, which is either constant or time-dependent. Many of these situations lead to closed-form solutions for a range of two-level problems. I include all the steps in the derivations for such problems, so the reader sees how the closed-form solution results. Numerous plots show how the solutions depend on the parameters of each problem and how the solutions evolve with time.

These closed-form solutions are joined by numerical solutions of the Schrödinger Equation with the aid of Mathematica. This allows us to go beyond the limited class of closed-form solutions. All lead to plots that illustrate the behavior of the solutions for the populations of the energy levels as functions of time. Experimental data are included and the solutions are used to extract information from the data.

I have pursued few-level problems because I find them interesting and because a single electron is often the source of the data. This has occurred more frequently as experimenters have become adept at isolating a single atom, or one molecule, or one defect. We are then watching Nature at a very basic level. The data often resemble random telegraph signals, which are also described and are fascinating to probe. My hope is that readers will find their time is profitably spent.

Chapter 1 introduces time dependence in Quantum Mechanics and shows how the solutions to the Schrödinger Equation evolve in time. These ideas are illustrated with the precession of the spin vector of a spin-$\frac{1}{2}$ particle. Then the Einstein coefficients and rate equations are introduced. The latter are often used as the first probe of a problem.

Chapter 2 treats two-level systems when the interaction is independent of time. The Sudden Approximation is discussed and an introduction is provided to random telegraph signals, which are also called random telegraph noise.

Time-dependent interactions enter in Chap. 3. The Rotating Wave Approximation is developed and its limits are explored. Closed-form solutions are found for a variety of monochromatic electromagnetic fields. The Rabi oscillations seen with a trapped ion are used to illustrate the closed-form solutions. Chapter 3's solutions are further applied in Chap. 4 to Rabi oscillations, the ammonia maser and magnetic resonance. Then the NV^- defect center in diamond is discussed, since this defect is an excellent example of a two-level defect in a solid.

Chapter 5 brings the density matrix to life for two-level systems. This formalism is useful for including relaxation phenomena. The Bloch Equations are derived and a solvable case is presented. This chapter concludes with another closed-form solution for the elements of the density matrix.

The second-order correlation function is developed in Chap. 6 for a two-level system in the context of resonance fluorescence. This function emerges from the density matrix and is numerically evaluated and compared to data from a quantum dot.

Several "Rambles" are included. These further explore topics that arise in the chapters. The book concludes with six appendices. The first provides values for the needed physical constants and the other five elucidate aspects of selected derivations.

There is "an elephant in the room", so to speak. The book does not engage with quantum computation nor with quantum information processing, although the material included in this book does help prepare a reader for such fields. I avoid these topics because there is a plethora of books devoted to these fields. I refer interested readers to:

1. Leslie E. Ballentine, *Quantum Mechanics: A Modern Development*, 2nd ed. (World Scientific, New Jersey and Singapore, 2015), Chap. 21.
2. Leonard Susskind and Art Friedman, *Quantum Mechanics: The Theoretical Minimum* (Basic Books, New York, 2014).
3. Eleanor Rieffel and Wolfgang Polak, *Quantum Computing: A Gentle Introduction* (MIT Press, Cambridge, MA, 2014).
4. Jochen Pade, *Quantum Mechanics for Pedestrians 2: Applications and Extensions* (Springer, Cham, 2014).
5. János A. Bergou and Mark Hillery, *Introduction to the Theory of Quantum Information Processing* (Springer Science + Business Media, New York, 2013).
6. Riley Tipton Perry, *Quantum Computing from the Ground Up* (World Scientific, Singapore, 2012).
7. Peter Lambropoulos and David Petrosyan, *Fundamentals of Quantum Optics and Quantum Information* (Springer-Verlag, Berlin, 2007).

I kept the derivations in this book close to their origins in the Time-Dependent Schrödinger Equation. Alternate approachs are common with the classic being

L. Allen and J. H. Eberly, *Optical Resonance and Two-Level Atoms*
(Dover Publications, New York, 1987).

Other books with significant coverage of time-dependent Quantum
Mechanics appear as references in the following chapters. My intent
is to supplement these texts rather than to supplant them. There are
also several excellent on-line sources for few-level problems. These
tend to be at a more advanced level than this text:

1. The MIT-Harvard Atomic Physics Wiki — https://cuax.mit.
   edu/apwiki/Main_Page
2. The MIT Open Course Ware has two graduate courses that
   overlap the Wiki:

   > Physics 8.421 Atomic and Optical Physics I
   > Physics 8.422 Atomic and Optical Physics II

3. Time-Dependent Quantum Mechanics and Spectroscopy by
   Andrei Tokmakoff — https://tdqms.uchicago.edu/
4. Quantum and Atom Optics by Daniel A. Steck — http://
   atomoptics-nas.uoregon.edu/~dsteck/teaching/quantum-optics/
   quantum-optics-notes.pdf

Many people kindly helped me understand the few-level material.
In particular, Thomas Gentile of the National Institute of Standards
and Technology corrected my erroneous ideas about his 1989 *Physical
Review A* paper. Michael D. Fayer of the Department of Chemistry of
Stanford University provided the details of his solution of a two-level
problem that appeared in his book *Elements of Quantum Mechanics*.
I am grateful to Mark Esrick of Georgetown University's Department
of Physics for his questions over the years as to my progress. The
Department of Physics has provided a home for the past several
years and I thank all the members for their help. I especially
thank James K. Freericks for his questions on the Rotating Wave
Approximation and Edward Van Keuren for his explanations of
Quantum Optics. I have also benefitted from the assistance of the
staff of Blommer Library, in particular, I appreciate the help of Jill
Hollingsworth. The late John Paul Shepherd and my late aunt, Lillian
Leffert, helped keep me focused. I thank my long-time friends and

colleagues Harvey S. Picker and Quang Ho-Kim for their support and their explanations. In addition, Quang has continually provided guidance on the writing of this book. I am grateful for the continuing friendship of my former colleagues at Eastman Kodak Company, Eric G. Stevens, William C. McColgin, and David D. Tuschel. They were always willing to listen and offer advice. I thank my cousin Jacqueline Rubenstein for her explanations of the publishing process. My wife Carolyn provided unquestioned encouragement and support throughout this project. And I thank her for reading the manuscript and smoothing its flow. If she checks the references in the following chapters, she will see I have used many of the Physics books that clutter our home. I thank Ng Kah-Fee and his colleagues at World Scientific Publishing Company for turning my manuscript into this book. I am grateful to the copy editor, Gregory Lee, for his many suggestions that improved my manuscript.

Notation used: vector $v = \bar{v}$

operator $H = \hat{H}$ (Please note that the hat is used in a few places to indicate a unit vector for a spatial coordinate.)

vector operator $S = \hat{\bar{S}}$

Hermitian adjoint of $M = M^{\dagger}$

The website www.quantumlavine.com has problems and solutions for two-level systems and the inevitable errata.

Contents

Chapter 1

Spin Precession and Rate Equations

Time dependence involves changes in the occupation of the states of a physical system. For example, the electron in a hydrogen atom may be excited from a 1s state into a 2p state due to a perturbation such as an electromagnetic field. (The states of the hydrogen atom are derived in Quantum Mechanics texts such as Konishi and Paffuti (2009, Chap. 6).) Generally, we say a transition has occurred when the physical system changes its state. It is simplest to view this as a change from the system being in state i to the system being in state j. This chapter starts with basic ideas on time dependence in Quantum Mechanics and the time evolution operator. These are applied to spin precession of a particle. Then the Einstein coefficients and rate equations are introduced. The following chapters are concerned with calculating the probability $p_i(t)$ for the occupation of state i of a system as a function of the time t. These predictions are compared with experiments whenever possible.

Section 1.1 briefly states what we explore and Sec. 1.2 introduces the Schrödinger Equation and the time evolution operator. These are applied to the precession of the spin angular momentum for a spin-$\frac{1}{2}$ particle in a constant magnetic field in Sec. 1.3. This example also serves as a review of the Pauli matrices. Section 1.4 presents the Einstein coefficients and the relationships between absorption, stimulated emission and spontaneous emission of electromagnetic radiation. Section 1.5 discusses rate equations, which are often used to describe transitions within a physical system. This chapter ends with two Rambles in Sec. 1.6. These touch on the wave function of the neutron and on the spin precession of the muon. Data are presented for both topics and compared with the theory developed in this chapter. Appendix 1 collects the values of useful physical constants, and App. 2 covers the electromagnetic field and includes some relationships that help us understand experiments.

1.1. Experiments and Time Dependence

Experimental probes take a variety of forms. We start with a beam of photons as an example of a probe of atoms by electromagnetic radiation. We assume a set of identical physical systems, each with its state 2 having a higher energy than its state 1. The photon beam falls on our sample and we measure the incident intensity and the transmitted intensity. From these we hope to deduce, for example, information on the absorption probability of state 1 versus the photon wavelength. Alternatively, we may observe our sample and detect the emission of radiation when a physical system makes a transition from state 2 to state 1. In both cases, we learn about the occupation probabilities for states 1 and 2 and the energy difference between them.

We let $p_1(t)$ and $p_2(t)$ be the occupation probabilities of states 1 and 2, respectively, as a function of the time t. These probabilities may be obtained from measurements on a sample with an ensemble of physical systems such as a volume of gas or of condensed matter. We need to measure $p_i(t)$ as a function of time. Then the transition probability for the physical system changing from state 2 to state 1 is

$$w_{12}(t) = \frac{dp_2(t)}{dt}, \qquad (1.1.1)$$

and for state 1 to state 2

$$w_{21}(t) = \frac{dp_1(t)}{dt}. \qquad (1.1.2)$$

The sample may also consist of an isolated atom, ion or molecule. In this case, the probabilities come from repeated measurements of the same physical system or from measurements of a beam of identical physical systems.

In order to understand the measured $p_i(t)$, we construct a model of a physical system and then solve the appropriate Schrödinger Equation. These solutions lead to predictions of $p_i(t)$ that we compare to the experimental results. This allows us to learn if our model system is realistic and, if so, to extract parameters of the physical system. In the following chapters we show how $p_i(t)$ is computed for a variety of physical systems and excitations or perturbations. This chapter continues with the introduction of the Time-Dependent Schrödinger Equation and the study of spin-$\frac{1}{2}$ precession.

1.2. Time Dependence and Time Evolution

We start with the time-dependent version of the Schrödinger Equation as presented in texts on Quantum Mechanics such as Cohen-Tannoudji *et al.* (1977), Konishi and Paffuti (2009), Sakurai and Napolitano (2011), Bransden and Joachain (2000) or Townsend (2000). We first find the time dependence of the wave function ψ for a time-independent Hamiltonian and then develop the time evolution operator.

Our wave function $\psi(\bar{x}, t)$ has a dependence on the spatial coordinates $\bar{x}$ and the time t. Here, $\bar{x}$ is a vector. The time dependence of the wave function is determined by solving the Schrödinger Equation

$$i\hbar \frac{\partial \psi(\bar{x}, t)}{\partial t} = \hat{H} \psi(\bar{x}, t). \tag{1.2.1}$$

The Hamiltonian operator is $\hat{H}$ and it **may** have an explicit dependence on the time. $\hbar$ is Planck's constant divided by 2π and i is the square root of -1.

We first assume the Hamiltonian $\hat{H}$ does **not** have any explicit dependence on time. So, we use the separation of variables and write

$$\psi(\bar{x}, t) = \phi(\bar{x})\theta(t). \tag{1.2.2}$$

This leads to

$$i\hbar \frac{d\theta(t)}{dt} = E\theta(t), \tag{1.2.3}$$

where E is the energy eigenvalue of the time-independent Hamiltonian $\hat{H}$, which comes from the eigenvalue equation

$$\hat{H}\phi(\bar{x}) = E\phi(\bar{x}). \tag{1.2.4}$$

This last equation is known as the time-independent Schrödinger Equation and it is the mainstay of most Quantum Mechanics texts. There are boundary conditions associated with Eq. (1.2.4) and normally a set of eigenvalues and eigenfunctions emerge as solutions. We use the index n for this set and solve Eq. (1.2.3) to find

$$\theta_n(t) = e^{-iE_nt/\hbar}. \tag{1.2.5}$$

We now write:

$$\psi(\bar{x}, t) = \sum_n a_n e^{-iE_nt/\hbar}\phi_n(\bar{x}), \tag{1.2.6}$$

and the set of constant coefficients $\{a_n\}$ is determined by the initial conditions for the problem.

We next construct an operator that advances the wave function in time,

$$\psi(\bar{x}, t) = \hat{U}(t, t_0)\psi(\bar{x}, t_0), \tag{1.2.7}$$

where the operator $\hat{U}$ is called the time evolution operator. Equation (1.2.6) suggests that for the special case of a time-independent Hamiltonian we try

$$\hat{U}(t, 0) = e^{-i\hat{H}t/\hbar}, \tag{1.2.8}$$

with $t_0 = 0$ for now. We expand the exponential of Eq. (1.2.8) to get a sum over powers of $\hat{H}$ and we easily evaluate each term, since the $\phi_n(\bar{x})$ are eigenfunctions of $\hat{H}$. For example,

$$\hat{H}^4\phi_n(\bar{x}) = (E_n)^4\phi_n(\bar{x}). \tag{1.2.9}$$

We then sum the resulting series to find

$$\hat{U}(t, 0)\phi_n(\bar{x}) = e^{-iE_nt/\hbar}\phi_n(\bar{x}), \tag{1.2.10}$$

and

$$\psi(\bar{x}, t) = e^{-i\hat{H}t/\hbar}\psi(\bar{x}, 0) = e^{-i\hat{H}t/\hbar}\sum_n a_n\phi_n(\bar{x})$$

$$= \sum_n a_n e^{-iE_n t/\hbar}\phi_n(\bar{x}). \tag{1.2.11}$$

This agrees with Eq. (1.2.6), so the suggested $\hat{U}$ of Eq. (1.2.8) works for this case.

The next step is to develop $\hat{U}$ more formally. Townsend (2000, pp. 93 ff.) and Sakurai and Napolitano (2011, pp. 66 ff.) are useful sources that supplement the present treatment. We concentrate on the time dependence of the wave function and develop the conditions $\hat{U}$ must meet. It is useful to shift to Dirac's bras and kets and to suppress the spatial dependence. Thus, the ket $|\theta, t_0; t\rangle$ satisfies the Schrödinger Equation with the Hamiltonian $\hat{H}$. The initial time is t_0 and the ket is evaluated at time t. The time evolution operator $\hat{U}$ takes the ket from t_0 to t,

$$|\theta, t_0; t\rangle = \hat{U}(t, t_0)|\theta, t_0; t_0\rangle. \tag{1.2.12}$$

We assume the ket at time t_0 is normalized

$$\langle\theta, t_0; t_0|\theta, t_0; t_0\rangle = 1, \tag{1.2.13}$$

and we assume particles are neither created nor destroyed. Thus, the time-evolved ket conserves probability and

$$\langle\theta, t_0; t_0|\theta, t_0; t_0\rangle = \langle\theta, t_0; t|\theta, t_0; t\rangle$$
$$= \langle\theta, t_0; t_0|\hat{U}^\dagger(t, t_0)\hat{U}(t, t_0)|\theta, t_0; t_0\rangle = 1. \tag{1.2.14}$$

This leads to the requirement that the time evolution operator is unitary

$$\hat{U}^\dagger(t, t_0)\hat{U}(t, t_0) = \hat{I}, \tag{1.2.15}$$

with $\hat{I}$ the identity operator. We also need $\hat{U}(t, t_0)$ to be the same as a sequence of time evolutions. For example,

$$\hat{U}(t_2, t_0) = \hat{U}(t_2, t_1)\hat{U}(t_1, t_0), \tag{1.2.16}$$

where $t_2 > t_1 > t_0$. And we can envision breaking the time interval down into even more time steps. This leads us to an additional condition on $\hat{U}$, namely that as t approaches t_0 we have

$$\hat{U}(t \to t_0, t_0) \to \hat{I}. \tag{1.2.17}$$

The above requirements on $\hat{U}$ are met to first-order in dt by

$$\hat{U}(t_0 + dt, t_0) = \hat{I} - \frac{i\hat{H}}{\hbar} dt, \tag{1.2.18}$$

with the Hamiltonian $\hat{H}$ Hermitian. Here $\hat{H}$ may be time dependent. This $\hat{U}$ satisfies Eqs. (1.2.15), (1.2.16) and (1.2.17) to first-order in dt as is verified by direct substitution. Now we use Eq. (1.2.16) with $t_2 = t + dt$ and $t_1 = t$ to see

$$\hat{U}(t + dt, t_0) = \hat{U}(t + dt, t)\hat{U}(t, t_0) = \left(\hat{I} - \frac{i\hat{H}}{\hbar} dt \right) \hat{U}(t, t_0),$$

$$\tag{1.2.19}$$

where we use the infinitesimal form of $\hat{U}$ from Eq. (1.2.18). Equation (1.2.19) is rearranged to

$$\hat{U}(t + dt, t_0) - \hat{U}(t, t_0) = \left(-\frac{i\hat{H}}{\hbar} dt \right) \hat{U}(t, t_0), \tag{1.2.20}$$

and in the limit of dt going to 0, we have

$$i\hbar \frac{\partial \hat{U}(t, t_0)}{\partial t} = \hat{H}\hat{U}(t, t_0). \tag{1.2.21}$$

This provides the Schrödinger Equation for the time evolution operator $\hat{U}(t, t_0)$ for the Hamiltonian $\hat{H}$. When we multiply Eq. (1.2.21) from the right by the ket $|\theta, t_0; t_0\rangle$, which is independent of the time t, and use Eq. (1.2.12), we find

$$i\hbar \frac{\partial |\theta, t_0; t\rangle}{\partial t} = \hat{H}|\theta, t_0; t\rangle, \tag{1.2.22}$$

the Schrödinger Equation for the ket $|\theta, t_0; t\rangle$. These developments allow us to tackle time-dependent Quantum Mechanics through the time evolution operator $\hat{U}$ or through the ket $|\theta, t_0; t\rangle$.

Of course, we still need to solve a differential equation. For the purposes of this chapter it suffices to treat a time-independent Hamiltonian $\hat{H}$. Then the formal solution to Eq. (1.2.21) is

$$\hat{U}(t, t_0) = e^{-i\hat{H}(t-t_0)/\hbar}. \tag{1.2.23}$$

If we expand the exponential in a power series we see that Eq. (1.2.21) is satisfied. In addition,

$$\hat{U}(t_0, t_0) = \hat{I}, \tag{1.2.24}$$

as needed. And Eq. (1.2.16) is

$$\hat{U}(t_2, t_0) = \hat{U}(t_2, t_1)\hat{U}(t_1, t_0) = e^{-i\hat{H}(t_2-t_1)/\hbar}e^{-i\hat{H}(t_1-t_0)/\hbar}$$

$$= e^{-i\hat{H}(t_2-t_0)/\hbar}, \tag{1.2.25}$$

when $\hat{H}$ is independent of time. The time evolution operator of Eq. (1.2.23) also gives the result of Eq. (1.2.8) with which we started this section. We will use this $\hat{U}$ for a time-independent Hamiltonian in the following discussion of spin precession.

1.3. Spin-$\frac{1}{2}$, the Pauli Matrices and Spin Precession

We now apply the time evolution operator $\hat{U}(t, t_0)$ of Eq. (1.2.23) to a time-independent Hamiltonian to investigate the free precession of a spin-$\frac{1}{2}$ particle. This precession provides the time dependence of the particle's spin vector. But first, we review the Pauli matrices since they are exceptionally useful in discussions of a spin-$\frac{1}{2}$ system. The Pauli matrices are introduced and developed in most Quantum Mechanics texts. We define the 2×2 Pauli matrices by

$$\sigma_1 = \begin{pmatrix} 0 & 1 \\ 1 & 0 \end{pmatrix}, \quad \sigma_2 = \begin{pmatrix} 0 & -i \\ i & 0 \end{pmatrix}, \quad \sigma_3 = \begin{pmatrix} 1 & 0 \\ 0 & -1 \end{pmatrix}. \tag{1.3.1}$$

We note that the subscripts 1, 2 and 3 are often replaced by x, y and z. Each Pauli matrix is Hermitian with its Trace, which is the sum of the diagonal matrix elements, equal to zero. In addition, the determinant of each Pauli matrix is equal to -1. Matrix multiplication shows

$$\sigma_i \sigma_i = \sigma_i^2 = I = \begin{pmatrix} 1 & 0 \\ 0 & 1 \end{pmatrix}, \tag{1.3.2}$$

where we introduce the 2×2 identity matrix I. Direct matrix multiplication also shows that for $i \neq j$

$$\sigma_i \sigma_j + \sigma_j \sigma_i = 0. \tag{1.3.3}$$

If we consider the vector $\bar{\sigma}$ to have the σ_i as components, then the dot product with the vector $\bar{v}$ becomes

$$\bar{\sigma} \cdot \bar{v} = \sum_{i=1}^{3} \sigma_i v_i = \begin{pmatrix} v_3 & v_1 - iv_2 \\ v_1 + iv_2 & -v_3 \end{pmatrix}. \tag{1.3.4}$$

We build on this to evaluate

$$(\bar{\sigma} \cdot \bar{v})^2 = (v_1^2 + v_2^2 + v_3^2)I = |\bar{v}|^2 I. \tag{1.3.5}$$

This follows from matrix multiplication and the square of the right-hand side of Eq. (1.3.4) or from writing all the cross terms that arise in squaring the sum in Eq. (1.3.4). The latter path draws on Eqs. (1.3.2) and (1.3.3).

For spin-$\frac{1}{2}$, we have the ket $|+\rangle$ for the spin up or z-component of spin $= +1/2$ and the ket $|-\rangle$ for the spin down with z-component $= -1/2$. We write these kets as column vectors

$$|+\rangle = \begin{pmatrix} 1 \\ 0 \end{pmatrix} \quad \text{and} \quad |-\rangle = \begin{pmatrix} 0 \\ 1 \end{pmatrix}. \tag{1.3.6}$$

The linear superposition is called a two-component spinor κ and we have

$$|\kappa\rangle = \begin{pmatrix} a_1 \\ a_2 \end{pmatrix} = a_1 |+\rangle + a_2 |-\rangle, \tag{1.3.7}$$

with the corresponding bra

$$\langle \kappa | = (a_1^* \quad a_2^*) = a_1^* \langle +| + a_2^* \langle -|. \tag{1.3.8}$$

We identify the spin operators $\hat{S}_i$, which are the components of the vector operator $\hat{\bar{S}}$, with

$$\hat{S}_i = \frac{\hbar}{2} \sigma_i. \tag{1.3.9}$$

We readily verify, using two-component spinors and the 2×2 matrix representation, that

$$\hat{S}_z|+\rangle = \frac{\hbar}{2}\begin{pmatrix} 1 & 0 \\ 0 & -1 \end{pmatrix}\begin{pmatrix} 1 \\ 0 \end{pmatrix} = \frac{\hbar}{2}\begin{pmatrix} 1 \\ 0 \end{pmatrix} = \frac{\hbar}{2}|+\rangle, \qquad (1.3.10\text{a})$$

and

$$\hat{S}_z|-\rangle = \frac{\hbar}{2}\begin{pmatrix} 1 & 0 \\ 0 & -1 \end{pmatrix}\begin{pmatrix} 0 \\ 1 \end{pmatrix} = \frac{\hbar}{2}\begin{pmatrix} 0 \\ -1 \end{pmatrix} = -\frac{\hbar}{2}\begin{pmatrix} 0 \\ 1 \end{pmatrix} = -\frac{\hbar}{2}|-\rangle. \qquad (1.3.10\text{b})$$

These results assure us that the above definitions are consistent.

The treatment of spin-$\frac{1}{2}$ precession requires the time evolution operator $\hat{U}$ of Eq. (1.2.23). This operator has the Hamiltonian in the exponential, and the simplest Hamiltonian for a spin interacting with a constant magnetic field $\bar{B}$ is

$$\hat{H} = -\hat{\bar{M}} \cdot \bar{B} = \frac{g\mu_B \hat{\bar{S}}}{\hbar} \cdot \bar{B}. \qquad (1.3.11)$$

Here $\hat{\bar{M}}$ is the magnetic dipole moment operator, g is the gyromagnetic ratio of the particle, which is taken to be an electron here, and μ_B is the Bohr magneton. Notation and definitions for this Hamiltonian vary from source to source. We follow Bransden and Joachain (2000). We need to evaluate

$$\hat{U}(t,0) = e^{-i\hat{H}t/\hbar} = e^{-i\frac{g\mu_B \hat{\bar{S}}}{\hbar}\cdot\bar{B}t/\hbar} = e^{-i\bar{\sigma}\cdot\bar{B}\zeta/2}, \qquad (1.3.12)$$

where we use Eq. (1.3.9) and set

$$\zeta = g\mu_B t/\hbar. \qquad (1.3.13)$$

The next step is to expand the exponential

$$\hat{U}(t,0) = e^{-i\bar{\sigma}\cdot\bar{B}\zeta/2} = I + \left(-i\frac{\bar{\sigma}\cdot\bar{B}\zeta}{2}\right) + \frac{1}{2!}\left(-i\frac{\bar{\sigma}\cdot\bar{B}\zeta}{2}\right)^2$$

$$+ \frac{1}{3!}\left(-i\frac{\bar{\sigma}\cdot\bar{B}\zeta}{2}\right)^3 + \cdots. \qquad (1.3.14)$$

We simplify this expansion through the use of

$$(\bar{\sigma}\cdot\bar{B})^2 = |\bar{B}|^2 I, \qquad (1.3.15)$$

and

$$(\bar{\sigma} \cdot \bar{B})^3 = (\bar{\sigma} \cdot \bar{B})(\bar{\sigma} \cdot \bar{B})^2 = (\bar{\sigma} \cdot \bar{B})|\bar{B}|^2, \qquad (1.3.16)$$

which follow from Eq. (1.3.5). We first group all the terms of Eq. (1.3.14) with even powers of $\bar{\sigma} \cdot \bar{B}$,

$$I\left[1 - \frac{1}{2!}\left(\frac{\zeta|\bar{B}|}{2}\right)^2 + \frac{1}{4!}\left(\frac{\zeta|\bar{B}|}{2}\right)^4 - \cdots\right] = I\cos(\zeta|\bar{B}|/2).$$
$$(1.3.17)$$

We next group the terms of Eq. (1.3.14) with odd powers of $\bar{\sigma} \cdot \bar{B}$

$$-i\left[\left(\frac{\bar{\sigma} \cdot \bar{B}}{|\bar{B}|}\right)\left(\frac{\zeta|\bar{B}|}{2}\right) - \left(\frac{\bar{\sigma} \cdot \bar{B}}{|\bar{B}|}\right)\left(\frac{\zeta|\bar{B}|}{2}\right)^3 + \cdots\right]$$
$$= -i\left(\frac{\bar{\sigma} \cdot \bar{B}}{|\bar{B}|}\right)\sin\left(\frac{\zeta|\bar{B}|}{2}\right). \qquad (1.3.18)$$

The last two equations lead to

$$\hat{U}(t,0) = e^{-i\bar{\sigma}\cdot\bar{B}\zeta/2} = I\cos\left(\frac{\zeta|\bar{B}|}{2}\right) - i\frac{(\bar{\sigma} \cdot \bar{B})}{|\bar{B}|}\sin\left(\frac{\zeta|\bar{B}|}{2}\right),$$
$$(1.3.19)$$

and the time t is hidden within ζ.

Often, the magnetic field $\bar{B}$ is such that

$$\bar{B} = B\bar{n}, \qquad (1.3.20)$$

with $\bar{n}$ a unit vector in the direction of the magnetic field of magnitude B. We regroup the constants and parameters to make the time explicit

$$\Omega = \frac{g\mu_B B}{\hbar}, \qquad (1.3.21)$$

and rewrite Eq. (1.3.19) as

$$\hat{U}(t,0) = e^{-i\bar{\sigma}\cdot\bar{n}B\zeta/2} = I\cos\left(\frac{\Omega t}{2}\right) - i(\bar{\sigma} \cdot \bar{n})\sin\left(\frac{\Omega t}{2}\right). \qquad (1.3.22)$$

We are finally ready for spin precession!

We consider a constant magnetic field in the z-direction and deduce how the electron's ket and the expectation values of the components of the spin angular momentum change with the time t. So the time evolution operator of Eq. (1.3.22) is now

$$\hat{U}(t,0) = e^{-i\sigma_z \Omega t/2} = I \cos\left(\frac{\Omega t}{2}\right) - i\sigma_z \sin\left(\frac{\Omega t}{2}\right). \qquad (1.3.23)$$

We write this as a 2×2 matrix

$$\hat{U}(t,0) = e^{-i\sigma_z \Omega t/2}$$

$$= \begin{pmatrix} \cos(\Omega t/2) - i\sin(\Omega t/2) & 0 \\ 0 & \cos(\Omega t/2) + i\sin(\Omega t/2) \end{pmatrix}$$

$$= \begin{pmatrix} e^{-i\Omega t/2} & 0 \\ 0 & e^{+i\Omega t/2} \end{pmatrix}. \qquad (1.3.24)$$

We use this time-evolution operator $\hat{U}$ to study two cases.

Our first case starts the electron in a spin-up state $|+\rangle$, which is an eigenket of the Hamiltonian in Eq. (1.3.23). We show how the phase of our ket evolves and then find the expectation values of the components of the spin vector. The ket at time zero with Eq. (1.3.7) is

$$|\kappa, 0; 0\rangle = \begin{pmatrix} 1 \\ 0 \end{pmatrix}, \qquad (1.3.25)$$

and at time t we have

$$|\kappa, 0; t\rangle = \hat{U}(t,0)|\kappa, 0; 0\rangle = \begin{pmatrix} e^{-i\Omega t/2} \\ 0 \end{pmatrix}. \qquad (1.3.26)$$

We note that when $\Omega t = 2\pi$, the phase of this ket is -1. It takes $\Omega t = 4\pi$ for the phase to return to $+1$. This phase behavior was verified for neutrons, which have spin $= \frac{1}{2}$, by Rauch *et al.* (1975) and Werner *et al.* (1975). The data of Rauch *et al.* (1975) are explored in Ramble 1 of Sec. 1.6. Further applications of neutron spin are discussed in a recent volume edited by Mezei (2010). In addition, Klempt (1976) provides data on the observation of the change of the phase factor for electrons from experiments with TlF molecules.

We return to expectation values with the aid of Eq. (1.3.9) and find with Eq. (1.3.26)

$$\langle \hat{S}_x \rangle = \frac{\hbar}{2} (e^{+i\Omega t/2} \quad 0) \begin{pmatrix} 0 & 1 \\ 1 & 0 \end{pmatrix} \begin{pmatrix} e^{-i\Omega t/2} \\ 0 \end{pmatrix}$$

$$= \frac{\hbar}{2} (e^{+i\Omega t/2} \quad 0) \begin{pmatrix} 0 \\ e^{-i\Omega t/2} \end{pmatrix} = 0, \qquad (1.3.27)$$

and similarly,

$$\langle \hat{S}_y \rangle = 0. \qquad (1.3.28)$$

Finally,

$$\langle \hat{S}_z \rangle = \frac{\hbar}{2} (e^{+i\Omega t/2} \quad 0) \begin{pmatrix} 1 & 0 \\ 0 & -1 \end{pmatrix} \begin{pmatrix} e^{-i\Omega t/2} \\ 0 \end{pmatrix}$$

$$= \frac{\hbar}{2} (e^{+i\Omega t/2} \quad 0) \begin{pmatrix} e^{-i\Omega t/2} \\ 0 \end{pmatrix} = \frac{\hbar}{2}. \qquad (1.3.29)$$

We see that when we start with an eigenket of $\hat{S}_z$, the expectation value of $\hat{S}_z$ does not change with time and the expectation values of the other two components are equal to zero. These results may be anticipated since we start with an eigenket of the time-independent Hamiltonian of Eq. (1.3.11).

However, for our second case, we verify that the expectation values of $\hat{S}_i$ have more variation when we start with a superposition of eigenkets of $\hat{S}_z$. Let us take

$$|\kappa, 0; 0\rangle = \frac{1}{\sqrt{2}} \begin{pmatrix} 1 \\ 1 \end{pmatrix}, \qquad (1.3.30)$$

which is actually an eigenket of $\hat{S}_x$ with the eigenvalue of $\hbar/2$,

$$\hat{S}_x |\kappa, 0; 0\rangle = \frac{1}{\sqrt{2}} \frac{\hbar}{2} \begin{pmatrix} 0 & 1 \\ 1 & 0 \end{pmatrix} \begin{pmatrix} 1 \\ 1 \end{pmatrix} = \frac{\hbar}{2} \frac{1}{\sqrt{2}} \begin{pmatrix} 1 \\ 1 \end{pmatrix} = \frac{\hbar}{2} |\kappa, 0; 0\rangle.$$

$$(1.3.31)$$

We now look at the time dependence. First,

$$|\kappa, 0; t\rangle = \hat{U}(t, 0)|\kappa, 0; 0\rangle = \begin{pmatrix} e^{-i\Omega t/2} & 0 \\ 0 & e^{+i\Omega t/2} \end{pmatrix} \frac{1}{\sqrt{2}} \begin{pmatrix} 1 \\ 1 \end{pmatrix}$$

$$= \frac{1}{\sqrt{2}} \begin{pmatrix} e^{-i\Omega t/2} \\ e^{+i\Omega t/2} \end{pmatrix}. \tag{1.3.32}$$

We note in passing that this time-evolved ket is no longer an eigenket of $\hat{S}_x$. We again use Eq. (1.3.9) to find

$$\langle \hat{S}_x(t) \rangle = \frac{\hbar}{4} (e^{+i\Omega t/2} \quad e^{-i\Omega t/2}) \begin{pmatrix} 0 & 1 \\ 1 & 0 \end{pmatrix} \begin{pmatrix} e^{-i\Omega t/2} \\ e^{+i\Omega t/2} \end{pmatrix}$$

$$= \frac{\hbar}{4} (e^{+i\Omega t/2} \quad e^{-i\Omega t/2}) \begin{pmatrix} e^{+i\Omega t/2} \\ e^{-i\Omega t/2} \end{pmatrix}$$

$$= \frac{\hbar}{4} (e^{i\Omega t} + e^{-i\Omega t}) = \frac{\hbar}{2} \cos \Omega t, \tag{1.3.33}$$

$$\langle \hat{S}_y(t) \rangle = \frac{\hbar}{4} (e^{+i\Omega t/2} \quad e^{-i\Omega t/2}) \begin{pmatrix} 0 & -i \\ i & 0 \end{pmatrix} \begin{pmatrix} e^{-i\Omega t/2} \\ e^{+i\Omega t/2} \end{pmatrix}$$

$$= \frac{\hbar}{4} (e^{+i\Omega t/2} \quad e^{-i\Omega t/2}) \begin{pmatrix} -ie^{+i\Omega t/2} \\ ie^{-i\Omega t/2} \end{pmatrix}$$

$$= \frac{\hbar}{2} \sin \Omega t, \tag{1.3.34}$$

and

$$\langle \hat{S}_z(t) \rangle = \frac{\hbar}{4} (e^{+i\Omega t/2} \quad e^{-i\Omega t/2}) \begin{pmatrix} 1 & 0 \\ 0 & -1 \end{pmatrix} \begin{pmatrix} e^{-i\Omega t/2} \\ e^{+i\Omega t/2} \end{pmatrix}$$

$$= \frac{\hbar}{4} (e^{+i\Omega t/2} \quad e^{-i\Omega t/2}) \begin{pmatrix} e^{-i\Omega t/2} \\ -e^{+i\Omega t/2} \end{pmatrix} = 0. \tag{1.3.35}$$

With the superposition of eigenkets of $\hat{S}_z$ we now have spin precession. The expectation values of $\hat{S}_x$ and $\hat{S}_y$ oscillate in time with the angular frequency of Ω defined by Eq. (1.3.21). They are periodic over a time interval of $2\pi/\Omega$. (This contrasts with the oscillations in the phase of the ket for a spin-$\frac{1}{2}$ particle, which require the

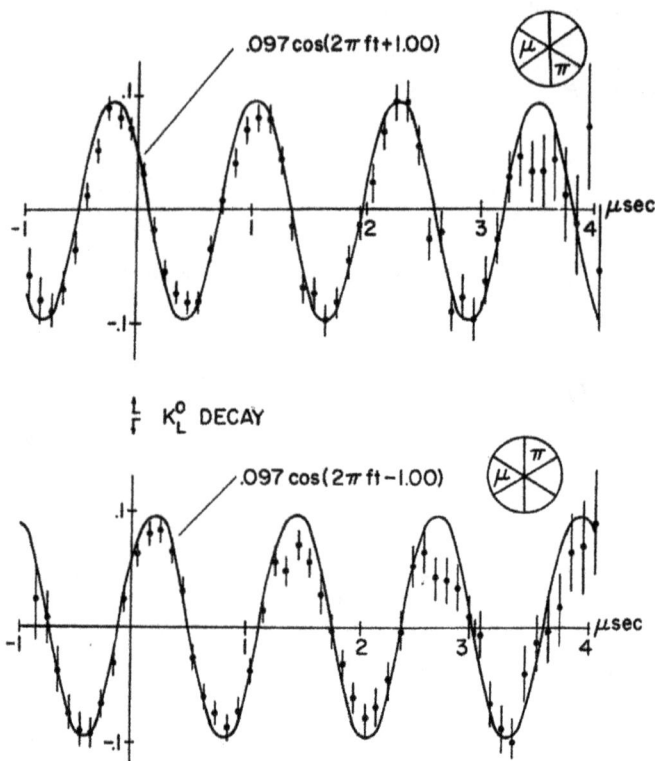

Figure 1.1. The projection of the muon spin vector versus time from Sandweiss *et al.* (1973). The oscillations follow the functional forms derived in Eqs. (1.3.33) and (1.3.34). Each set of data is for a separate detector. Please see Ramble 2 in Sec. 1.6. Reprinted figure with permission from Sandweiss *et al.* (1973). Copyright 1973 by the American Physical Society.

time interval of $4\pi/\Omega$, as seen in Eq. (1.3.26).) This behavior holds for other spin-$\frac{1}{2}$ particles such as the muon and the neutron. The challenge is to track the precession through an experiment. Sandweiss *et al.* (1973) show their muon spin precession results, which are reproduced here as Fig. 1.1 and discussed further in Ramble 2 at the end of this chapter.

It is natural to ask about data for the spin precession of the electron in a magnetic field. The challenge comes from moving electrons undergoing cyclotron motion at a frequency, the Larmor frequency, which is extremely close to the spin precession frequency. Very careful experiments have achieved a separation of these two frequencies (Wilkinson and Crane, 1963). Much recent research has focused on the transport of electron spins for spintronics and electron spin precession has been observed in materials such as graphene (Tombros *et al.*, 2007).

We round out the present discussion of time dependence with an observation on the time dependence of the expectation value of an operator $\hat{G}$. We return to the Schrödinger Equation with Hamiltonian $\hat{H}$ as given in Eq. (1.2.22) and use it and its adjoint to evaluate

$$\frac{d\langle \hat{G}(t)\rangle}{dt} = \frac{d}{dt}\langle \theta, t_0; t|\hat{G}(t)|\theta, t_0; t\rangle, \qquad (1.3.36)$$

and a commutator eventually appears

$$\frac{d\langle \hat{G}(t)\rangle}{dt} = \left(\frac{d}{dt}\langle \theta, t_0; t|\right)\hat{G}(t)|\theta, t_0; t\rangle + \left\langle \theta, t_0; t\left|\frac{\partial \hat{G}(t)}{\partial t}\right|\theta, t_0; t\right\rangle$$

$$+ \langle \theta, t_0; t|\hat{G}(t)\left(\frac{d}{dt}|\theta, t_0; t\rangle\right)$$

$$= \left(\frac{1}{-i\hbar}\langle \theta, t_0; t|\hat{H}\right)\hat{G}(t)|\theta, t_0; t\rangle + \left\langle \theta, t_0; t\left|\frac{\partial \hat{G}(t)}{\partial t}\right|\theta, t_0; t\right\rangle$$

$$+ \langle \theta, t_0; t|\hat{G}(t)\left(\frac{1}{i\hbar}\hat{H}|\theta, t_0; t\rangle\right)$$

$$= \frac{i}{\hbar}\langle \theta, t_0; t|[\hat{H}, \hat{G}]|\theta, t_0; t\rangle + \left\langle \theta, t_0; t\left|\frac{\partial \hat{G}(t)}{\partial t}\right|\theta, t_0; t\right\rangle.$$

$$(1.3.37)$$

The last term is needed when the operator $\hat{G}$ has an explicit dependence on time. We return to our assumption that $\hat{H}$ is proportional to $\hat{S}_z = \hbar\sigma_z/2$ and we consider $\hat{G}$ to be an $\hat{S}_i$. Then we have $\hat{G}$ explicitly independent of time, so

$$\frac{d\langle \hat{G}\rangle}{dt} = \frac{i}{\hbar}\langle \theta, t_0; t|[\hat{H}, \hat{G}]|\theta, t_0; t\rangle. \qquad (1.3.38)$$

When $\hat{G} = \hat{S}_z$ the commutator in Eq. (1.3.38) is $[\hat{S}_z, \hat{S}_z] = 0$ and the expectation value is a constant as we found in Eq. (1.3.35).

When $\hat{G} = \hat{S}_x$ or $\hat{G} = \hat{S}_y$, we find Eq. (1.3.38) and Eq. (1.3.32), which has $t_0 = 0$, lead to

$$\frac{d\langle \hat{S}_x(t) \rangle}{dt} = -\frac{\hbar}{2}\Omega \sin \Omega t, \qquad (1.3.39)$$

and

$$\frac{d\langle \hat{S}_y(t) \rangle}{dt} = \frac{\hbar}{2}\Omega \cos \Omega t, \qquad (1.3.40)$$

respectively. We note these results are consistent with Eqs. (1.3.33) and (1.3.34).

Next we discuss the Einstein coefficients and rate equations. Both topics are useful in understanding experiments.

1.4. The Einstein Coefficients

The chapters that follow are concerned with transitions between quantum states induced by electromagnetic fields. Thus, it is useful to review the work of Planck and Einstein, which led to the idea that electromagnetic radiation is absorbed or emitted when an atom or molecule changes its quantum state. For example, an electron bound to an atom goes from its present state in energy level 1 to a higher energy state, energy level 2, upon the absorption of a photon of energy $\hbar\omega$. Then

$$E_2 = E_1 + \hbar\omega, \qquad (1.4.1)$$

where E_i is the energy of energy level i. A transition of the electron from energy level 2 to the lower energy level 1 is accompanied by the emission of a photon of energy

$$\hbar\omega = E_2 - E_1. \qquad (1.4.2)$$

Now the atom needs to be in an electromagnetic field for the absorption process to take place. However, spontaneous emission may occur in the absence of the electromagnetic field. Einstein deduced that stimulated emission also takes place when an electromagnetic

Change in Energy Level Occupation by

E_2 ———
E_1 ———
Absorption in EM field

E_2 ———
E_1 ———
Stimulated emission in EM field

E_2 ———
E_1 ———
Spontaneous emission

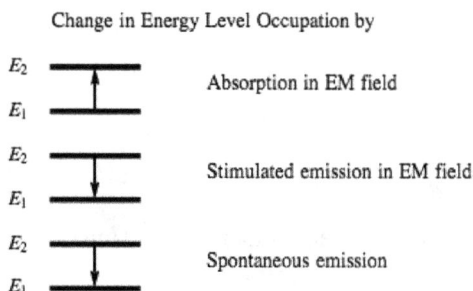

Figure 1.2. The transitions between two energy levels either due to an electromagnetic field or to spontaneous emission.

field is present. All three processes are illustrated in Fig. 1.2. In this chapter, we find the relations between these processes and then solve the resulting rate equations.

Our approach parallels Bransden and Joachain (2000, Sec. 11.3). We note that some authors interchange the order of the subscripts for the terms introduced below. We consider a set of identical atoms and a specific atomic transition between energy levels 1 and 2 with energies E_1 and E_2, respectively. We take energy level 2 to be the higher energy level. Let us assume that at time $t = 0$ there are $N_1(0)$ atoms in state 1 and $N_2(0)$ atoms in state 2. We next relate the time derivatives of $N_i(t)$ to the Einstein A and B coefficients (Einstein, 1917), which we define next. The population $N_2(t)$ increases by absorption and decreases by emission. We let $N_{21}(t)$ be the number of atoms promoted or excited from state 1 to state 2. This is also the number of upward transitions or absorption events. Then

$$\frac{dN_{21}(t)}{dt} = B_{21}N_1(t)\tilde{\rho}(\omega_{21}), \qquad (1.4.3)$$

with

$$\omega_{21} = (E_2 - E_1)/\hbar. \qquad (1.4.4)$$

B_{21} is the Einstein coefficient for absorption from state 1 to state 2 and $\tilde{\rho}(\omega_{21})$ is the energy density of the electromagnetic field per unit angular frequency at ω_{21}. The quantity $\tilde{\rho}(\omega_{21})$ is used because the energy density needed here has units different from the volume energy density $\rho(\omega)$ of App. 2. Similarly, $N_{12}(t)$ is the number of atoms that go from state 2 to state 1 by an emission event, which is a downward transition. Now

$$\frac{dN_{12}(t)}{dt} = A_{12}N_2(t) + B_{12}N_2(t)\tilde{\rho}(\omega_{21}). \qquad (1.4.5)$$

Here A_{12} is the Einstein coefficient for spontaneous emission and B_{12} is the Einstein coefficient for stimulated emission.

The three Einstein coefficients are related. We assume our set of atoms is in a box and the set of atoms is in thermal equilibrium with the walls of the box at temperature T. So we have thermal equilibrium between the atoms and the radiation in the box. Then the rates of absorption and emission are equal,

$$\frac{dN_{21}(t)}{dt} = \frac{dN_{12}(t)}{dt}. \qquad (1.4.6)$$

Thus, in thermal equilibrium,

$$B_{21}N_1\tilde{\rho}(\omega_{21}) = A_{12}N_2 + B_{12}N_2\tilde{\rho}(\omega_{21}), \qquad (1.4.7)$$

or

$$\frac{N_1}{N_2} = \frac{A_{12} + B_{12}\tilde{\rho}(\omega_{21})}{B_{21}\tilde{\rho}(\omega_{21})}. \qquad (1.4.8)$$

Equation (1.4.7) also leads to

$$\tilde{\rho}(\omega_{21}) = \frac{A_{12}N_2}{B_{21}N_1 - B_{12}N_2} = \frac{A_{12}/B_{12}}{\frac{B_{21}N_1}{B_{12}N_2} - 1}. \qquad (1.4.9)$$

We now recall the Boltzmann energy factor for a system in thermal equilibrium at temperature T and Baierlein (1999, Sec. 5.2) covers this nicely. The result is

$$\frac{N_2}{N_1} = \frac{g_2}{g_1}e^{-\hbar\omega_{21}/k_B T}, \qquad (1.4.10)$$

where k_B is Boltzmann's constant and g_i is the statistical weight or the degeneracy factor for state i. That is, we count the number of states with the same energy. We also need Planck's formula for the equilibrium distribution of blackbody radiation with frequency. This is found in Baierlein (1999, Sec. 6.2) and Bransden and Joachain (2000, p. 578). Planck found

$$\tilde{\rho}(\omega_{21}) = \frac{\hbar\omega_{21}^3}{\pi^2 c^3} \frac{1}{e^{\hbar\omega_{21}/k_B T} - 1}, \tag{1.4.11}$$

with c the speed of light. We require that Eq. (1.4.9) agrees with Eq. (1.4.11) when we insert Eq. (1.4.10) and this leads to

$$\frac{B_{21} g_1}{B_{12} g_2} = 1, \tag{1.4.12}$$

and

$$\frac{A_{12}}{B_{12}} = \frac{\hbar\omega_{21}^3}{\pi^2 c^3}. \tag{1.4.13}$$

Thus, the coefficients for absorption and stimulated emission are related through Eq. (1.4.12). In addition, if experiment or theory provides A_{12}, the coefficient for spontaneous emission, then Eqs. (1.4.13) and (1.4.12) yield the Einstein B_{12} and B_{21} coefficients, respectively. We note that the Einstein coefficients have been derived in thermal equilibrium. They are also useful guides in non-equilibrium situations.

Equation (1.4.13) shows the importance of spontaneous emission decreases as the frequency decreases. This is apparent from

$$\frac{A_{12}}{B_{12}} = \frac{\hbar\omega_{21}^3}{\pi^2 c^3} = \frac{4h\nu_{21}^3}{c^3} = \frac{4h\nu_{21}^3}{(\nu_{21}\lambda_{21})^3} = \frac{4h}{\lambda_{21}^3}, \tag{1.4.14}$$

with λ_{21} the wavelength corresponding to the angular frequency ω_{21}. Thus, the ratio in Eq. (1.4.14) drops by 1,000 for each tenfold increase in the wavelength. Hence, we may usually neglect A_{12} with respect to B_{12}. We confirm this by examining the ratio R of the rate of spontaneous emission to the rate of stimulated emission. We use the terms on the right-hand side of Eq. (1.4.5) to see

$$R = \frac{A_{12} N_2(t)}{B_{12}\tilde{\rho}(\omega_{21}) N_2(t)} = \frac{A_{12}}{B_{12}\tilde{\rho}(\omega_{21})} = e^{\hbar\omega_{21}/k_B T} - 1, \tag{1.4.15}$$

with the help of Eqs. (1.4.11) and (1.4.13). Figure 1.3 shows R versus the transition energy for the temperature $T = 300$ K, which is 0.025858 eV. We see that in the microwave region with $\lambda_{21} \sim 10^{-2}$ m,

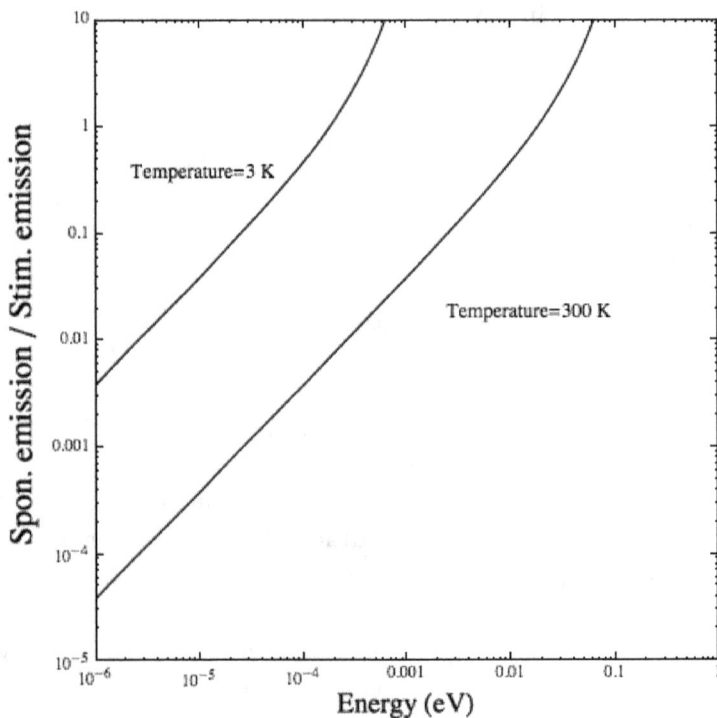

Figure 1.3. The ratio of spontaneous emission to stimulated emission from Eq. (1.4.15). The ratio is plotted for temperatures of 3 K and 300 K versus the photon energy in eV.

or $E_{21} \sim 1.24 \times 10^{-4}$ eV, and for longer λ_{21}, spontaneous emission is usually ignored. We add the ratio R for 3 K to Fig. 1.3. It is apparent that spontaneous emission becomes more important at a fixed E_{21} when the temperature decreases.

Section 1.5 continues with rate equations.

1.5. Rate Equations

We next derive Eqs. (1.4.12) and (1.4.13) from a direct consideration of the populations of energy levels 1 and 2 through rate equations. These yield the populations as a function of time and are discussed in Rigamonti and Carretta (2009, pp. 48–49). We use the notation introduced in Sec. 1.4. These rate equations also provide the long-time behavior of the two populations. For each energy level, we write a rate equation for how the number of atoms in the energy level changes with time. Thus,

$$\frac{dN_1(t)}{dt} = -B_{21}\tilde{\rho}(\omega_{21})N_1(t) + B_{12}\tilde{\rho}(\omega_{21})N_2(t) + A_{12}N_2(t),$$

$$(1.5.1)$$

$$\frac{dN_2(t)}{dt} = +B_{21}\tilde{\rho}(\omega_{21})N_1(t) - B_{12}\tilde{\rho}(\omega_{21})N_2(t) - A_{12}N_2(t).$$

$$(1.5.2)$$

These equations show how the populations of the two energy levels change by the absorption and emission of electromagnetic radiation of energy $\hbar\omega_{21}$. Equation (1.5.1) says $N_1(t)$ decreases through the first term on the right, which represents the absorption of a photon of energy $\hbar\omega_{21}$. The second and third terms on the right say $N_1(t)$ increases through stimulated and spontaneous emission, respectively, from energy level 2. Equation (1.5.2) is interpreted in a parallel fashion. We note that adding Eqs. (1.5.1) and (1.5.2) leads to

$$\frac{d}{dt}[N_1(t) + N_2(t)] = \frac{dN}{dt} = 0, \qquad (1.5.3)$$

and defines N, the total number of atoms. Equation (1.5.3) reassures us that N is conserved.

In thermal equilibrium, the population of each level is constant, so we have

$$\frac{d}{dt}[N_1(t)] = \frac{d}{dt}[N_2(t)] = 0. \qquad (1.5.4)$$

Both equations lead to,

$$B_{21}\tilde{\rho}(\omega_{21})N_1 = B_{12}\tilde{\rho}(\omega_{21})N_2 + A_{12}N_2, \qquad (1.5.5)$$

which is Eq. (1.4.7). Hence, this approach leads to the same relationships between A_{12}, B_{21} and B_{12}.

We now solve Eq. (1.5.1) to get N_1 as a function of time and to see the long-time behavior. We use

$$N = N_1(t) + N_2(t), \qquad (1.5.6)$$

with N a given constant, to remove $N_2(t)$. This last equation then gives us $N_2(t)$ after we have determined $N_1(t)$. We also let

$$W_{ij} = B_{ij}\tilde{\rho}(\omega_{21}), \qquad (1.5.7)$$

to simplify the notation. In accord with Eq. (1.4.12), we allow $g_1 \neq g_2$, so $W_{12} \neq W_{21}$. Equation (1.5.1) becomes

$$\frac{dN_1(t)}{dt} = -N_1(t)(W_{21} + W_{12} + A_{12}) + N(W_{12} + A_{12}), \qquad (1.5.8)$$

which is a first-order ordinary differential equation for $N_1(t)$. We try

$$N_1(t) = C + De^{\gamma t}, \qquad (1.5.9)$$

with the constants C and D. We substitute our trial solution into Eq. (1.5.8) and find

$$D\gamma e^{\gamma t} = -(C + De^{\gamma t})(W_{21} + W_{12} + A_{12}) + N(W_{12} + A_{12}), \qquad (1.5.10)$$

The terms with the exponential in the time give

$$\gamma = -(W_{21} + W_{12} + A_{12}), \qquad (1.5.11)$$

and the constant terms yield

$$C = N(W_{12} + A_{12})/(W_{21} + W_{12} + A_{12}). \qquad (1.5.12)$$

Finally, the constant D is set by the initial condition $N_1(t = 0)$.

We explore our solution and start with all the atoms in the lower energy state, so

$$N_1(0) = N, \qquad (1.5.13)$$
$$N_2(0) = 0. \qquad (1.5.14)$$

These lead to

$$D = NW_{21}/(W_{21} + W_{12} + A_{12}), \qquad (1.5.15)$$

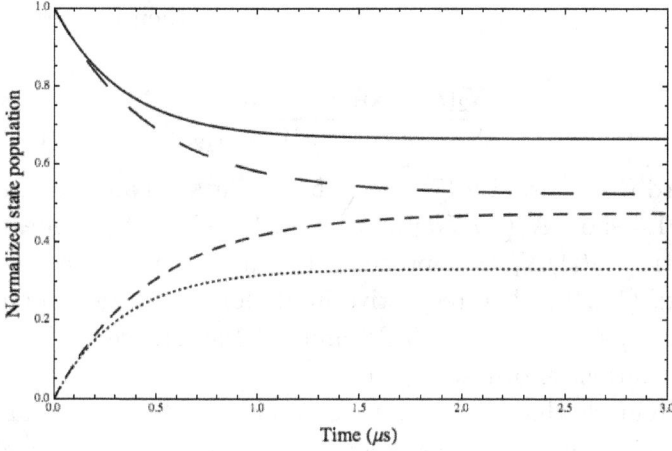

Figure 1.4. The normalized state populations $N_1(t)/N$ (solid) and $N_2(t)/N$ (dotted) for $W_{12} = W_{21} = 1.0/\mu$s and $A_{12} = 1.0/\mu$s and $N_1(t)/N$ (long dashes) and $N_2(t)/N$ (short dashes) for $W_{12} = W_{21} = 1.0/\mu$s and $A_{12} = 0.1/\mu$s. At $t = 0$, $N_1 = N$ and $N_2 = 0$.

and

$$N_1(t) = \frac{N[(W_{12} + A_{12}) + W_{21}e^{-(W_{21}+W_{12}+A_{12})t}]}{(W_{21} + W_{12} + A_{12})}. \qquad (1.5.16)$$

We plot $N_1(t)/N$ and $N_2(t)/N$ for $W_{12} = W_{21} = 1.0/\mu$s and $A_{12} = 1.0/\mu$s (solid and dotted curves) and $0.1/\mu$s (dashed curves) in Fig. 1.4. We see that $N_1(t)$ and $N_2(t)$ approach steady-state values, which we next determine.

At long times,

$$N_1(t \to \infty) \to \frac{N(W_{12} + A_{12})}{(W_{21} + W_{12} + A_{12})}, \qquad (1.5.17)$$

$$N_2(t \to \infty) = [N - N_1(t \to \infty)] \to \frac{NW_{21}}{(W_{21} + W_{12} + A_{12})}. \qquad (1.5.18)$$

Their ratio approaches

$$\frac{N_2(t \to \infty)}{N_1(t \to \infty)} \to \frac{W_{21}}{W_{12} + A_{12}}. \qquad (1.5.19)$$

If we set the degeneracy factors equal, $g_1 = g_2$, then $W = W_{12} = W_{21}$ and

$$\frac{N_2(t \to \infty)}{N_1(t \to \infty)} \to \frac{W}{W + A_{12}}. \tag{1.5.20}$$

In accord with Eqs. (1.5.17) and (1.5.18), the solid and dotted curves in Fig. 1.4 show $N_1(t)/N$ approaches 2/3 and $N_2(t)/N$ goes to 1/3. The ratio $N_2(t)/N_1(t)$ approaches 0.5 at long times in agreement with Eq. (1.5.19). The respective limits for the dashed curves with $A_{12} = 0.1/\mu s$ are 1.1/2.1, 1.0/2.1 and 1/1.1 and these are in line with the calculations shown in Fig. 1.4.

We remark that $N_2 < N_1$ in general and they are equal if we can neglect the spontaneous emission term A_{12}. This means that we cannot readily achieve the population inversion, or $N_2 > N_1$, needed for maser and laser operation. This explains why such devices almost always use three- or more level systems as discussed in Wilson and Hawkes (1989, Sec. 5.4) and in other books on optoelectronics. However, we do consider the first maser in Chap. 4, since it was a two-level system based on molecular beams of ammonia. The use of molecular beams avoids the constraint of Eq. (1.5.20) since a fresh supply of ammonia molecules in the higher-energy state is continually introduced.

For completeness, we consider Eq. (1.5.9) and

$$N_1(t = 0) = 0, \tag{1.5.21}$$

$$N_2(t = 0) = N, \tag{1.5.22}$$

which lead to

$$D = -\frac{N(W_{12} + A_{12})}{(W_{21} + W_{12} + A_{12})}. \tag{1.5.23}$$

The solutions are now

$$N_1(t) = \frac{N[(W_{12} + A_{12}) - (W_{12} + A_{12})e^{-(W_{21}+W_{12}+A_{12})t}]}{(W_{21} + W_{12} + A_{12})}, \tag{1.5.24}$$

$$N_2(t) = \frac{N[W_{21} + (W_{12} + A_{12})e^{-(W_{21}+W_{12}+A_{12})t}]}{(W_{21} + W_{12} + A_{12})}. \tag{1.5.25}$$

We again find for long times

$$\frac{N_2(t \to \infty)}{N_1(t \to \infty)} \to \frac{W_{21}}{W_{12} + A_{12}}, \qquad (1.5.26)$$

and we note that starting all the atoms in the higher-energy state does not help us achieve population inversion. We plot $N_1(t)/N$ and $N_2(t)/N$ for $W_{12} = W_{21} = 1.0/\mu s$ and $A_{12} = 1.0/\mu s$ (solid and dotted curves) and $0.1/\mu s$ (dashed curves) in Fig. 1.5. We see the populations approach the long-time limits found in Fig. 1.4 once the population of level 2 falls below that of level 1.

The rate equation approach of this section avoids coherent effects and to incorporate them we turn to a study of the Time-Dependent Schrödinger Equation. We start with two-level systems in Chaps. 2 and 3 and find the time development for a variety of excitation mechanisms. But first we delve further into spin precession with Rambles 1 and 2.

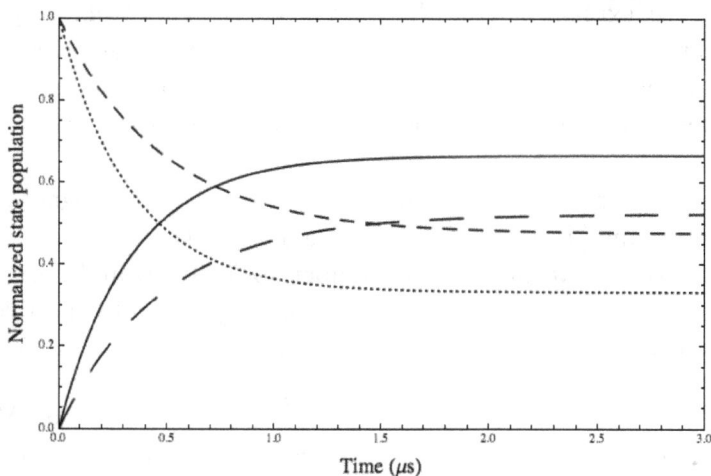

Figure 1.5. The normalized state populations $N_1(t)/N$ (solid) and $N_2(t)/N$ (dotted) for $W_{12} = W_{21} = 1.0/\mu s$ and $A_{12} = 1.0/\mu s$ and $N_1(t)/N$ (long dashes) and $N_2(t)/N$ (short dashes) for $W_{12} = W_{21} = 1.0/\mu s$ and $A_{12} = 0.1/\mu s$. At $t = 0$, $N_1 = 0$ and $N_2 = N$.

1.6. Two Rambles on Spin Precession for Spin-$\frac{1}{2}$ Particles

These two Rambles describe experiments that provide data for the phase of the neutron's wave function and for the spin precession of the muon. We show that the data agree with the theory developed in Sec. 1.3. Those seeking details beyond the following discussions may find it profitable to consult the cited references or texts on Quantum Mechanics, such as Robinett (2006) and Townsend (2000). The physical constants are taken from App. 1, although fewer significant digits are used here.

Ramble 1 — The neutron phase shift in a magnetic field

The ket for a spin-$\frac{1}{2}$ particle has a phase that goes from $+1$ to -1 when its argument advances from 0 to 2π. Another 2π is required for the phase to return to $+1$. Equation (1.3.26) points this out. The following experiments use neutron interference and are reported in Rauch *et al.* (1975) and Werner *et al.* (1975). A further experiment is detailed in Badurek *et al.* (1976). We now give a qualitative explanation of the data from Rauch *et al.*. A perfect crystal neutron interferometer yields two coherent beams of neutrons that are spatially separated as in Fig. 1.6. One beam passes through a region with a magnetic field present and this introduces a phase shift between the two beams.

First, we treat the neutrons that pass through a magnetic field $\bar{B}$. We assume $\bar{B}$ is in the z-direction and the neutron beam is unpolarized, so we have equal numbers of spin up and spin down neutrons. Let us first consider a constant magnetic field and take the Hamiltonian to be

$$\hat{H} = -\hat{\bar{M}} \cdot \bar{B}. \tag{1.6.1}$$

Here the magnetic moment operator for the neutron follows Robinett (2006, p. 480) and is

$$\hat{\bar{M}} = -\frac{g_n}{2}\mu_N\bar{\sigma}, \tag{1.6.2}$$

with the neutron's gyromagnetic ratio g_n equal to 2 times -1.91304 or -3.82608, μ_N the magnetic moment of the proton and $\bar{\sigma}$ the vector

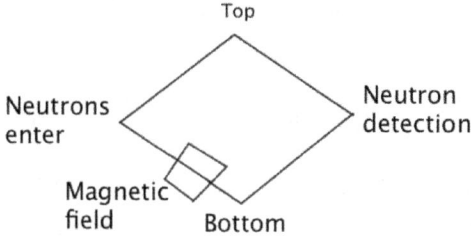

Figure 1.6. Schematic of the top and bottom neutron paths in the experiment of Rauch *et al.* (1975). The magnetic field is perpendicular to the plane of the page.

of the Pauli spin matrices. μ_N is not zero because the neutron is a composite particle. We have

$$\mu_N = \frac{e\hbar}{2m_p} = \frac{1.602176 \times 10^{-19}(1.05457 \times 10^{-34})}{2(1.67262 \times 10^{-27})} \text{ J/T}$$

$$= 5.05078 \times 10^{-27} \text{ J/T}, \qquad (1.6.3)$$

and m_p is the proton mass. Next, we look at the neutron paths.

The neutron beam is split into two and each neutron follows either the top leg or the bottom leg of Fig. 1.6. The bottom leg may contain a region of magnetic field in the z-direction perpendicular to the plane of the legs. The neutrons from the two legs are combined and detected. We assume the velocity of the neutron remains the same throughout its journey from injection to detection. The Hamiltonian for a free particle then introduces a phase factor for the time evolution operator according to Eqs. (1.2.7) to (1.2.11). Since we assume this phase factor is the same for every neutron, we set the phase factor to 1.0. We calculate the ratio of the observed intensities, with and without a magnetic field in the bottom leg. This ratio is proportional to

$$\frac{\||\psi_{top}(\bar{B} \neq 0)\rangle + |\psi_{bot}(\bar{B} \neq 0)\rangle\|^2}{\||\psi_{top}(\bar{B} = 0)\rangle + |\psi_{bot}(\bar{B} = 0)\rangle\|^2}. \qquad (1.6.4)$$

When the magnetic field $\bar{B} = 0$, we have

$$|\psi_{top}\rangle = |\psi_{bot}\rangle = \frac{1}{\sqrt{2}} \begin{pmatrix} 1 \\ 1 \end{pmatrix}. \qquad (1.6.5)$$

With the magnetic field in the z-direction, the Hamiltonian of Eq. (1.6.1) becomes

$$\hat{H} = -\frac{g_n}{2}\mu_N\sigma_z B_z = -\frac{g_n}{2}\mu_N B_z \begin{pmatrix} 1 & 0 \\ 0 & -1 \end{pmatrix}, \tag{1.6.6}$$

and the Pauli matrix is revealed. The magnetic field is not constant along the bottom leg of Fig. 1.6, but we use the integral over time of the magnetic field the neutrons pass through. Basically, we replace the Ωt in Eq. (1.3.24) with the spin precession angle θ that we now define through

$$d\theta = \omega dt = \frac{g_n}{\hbar}\mu_N B_z dt. \tag{1.6.7}$$

Here ω is the angular frequency of the spin precession. We have assumed the neutron has the constant velocity v and this allows us to change from a time integral to a spatial integral along the path with the coordinate s,

$$\theta = \int d\theta = \frac{g_n}{\hbar v}\mu_N \int B_z ds. \tag{1.6.8}$$

This integral over the neutron's path is measured during the experiment of Rauch *et al.* (1975) and the velocity is obtained from the neutron's wavelength λ_n of $1.82\pm0.01\times10^{-10}$ m and the neutron's mass m_n. This results in

$$v = \frac{h}{m_n\lambda_n} = \frac{6.626059 \times 10^{-34}}{1.674927 \times 10^{-27}(1.82 \times 10^{-10})} \text{ m/s}$$

$$= 2.1736 \times 10^3 \text{ m/s}. \tag{1.6.9}$$

We have all the pieces for the kets with a non-zero magnetic field. We write them in terms of the angle θ by analogy with Eq. (1.3.24) for the time evolution operator

$$|\psi_{bot}(\theta)\rangle = \frac{1}{\sqrt{2}}\begin{pmatrix} e^{-i\theta/2} & 0 \\ 0 & e^{i\theta/2} \end{pmatrix}\begin{pmatrix} 1 \\ 1 \end{pmatrix} = \frac{1}{\sqrt{2}}\begin{pmatrix} e^{-i\theta/2} \\ e^{i\theta/2} \end{pmatrix}, \tag{1.6.10}$$

and

$$|\psi_{top}(\theta)\rangle = \frac{1}{\sqrt{2}}\begin{pmatrix} 1 \\ 1 \end{pmatrix}. \tag{1.6.11}$$

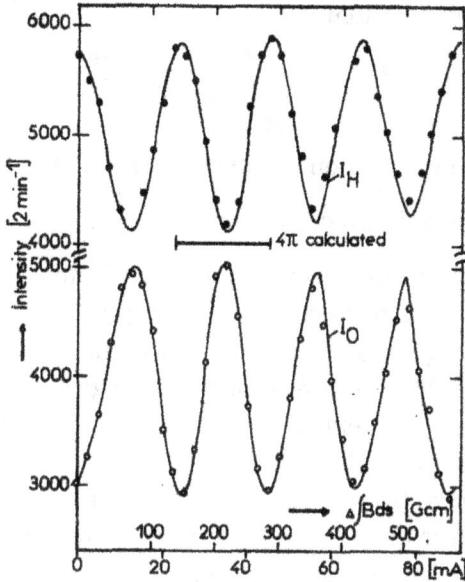

Figure 1.7. The neutron intensity oscillations observed by Rauch *et al.* (1975) versus the integral of the magnetic field over the path of the neutrons. The interval corresponding to 4π is marked. Reprinted figure from Rauch *et al.* (1975) with permission from Elsevier.

We substitute Eqs. (1.6.10) and (1.6.11) into the numerator of Eq. (1.6.4) with its denominator coming from Eq. (1.6.5). The result is

$$\frac{\||\psi_{top}(\theta)\rangle + |\psi_{bot}(\theta)\rangle|^2}{\||\psi_{top}(\theta=0)\rangle + |\psi_{bot}(\theta=0)\rangle|^2} = \frac{2(1+\cos(\theta/2))}{2(2)}$$

$$= \frac{1}{2}(1+\cos(\theta/2)). \quad (1.6.12)$$

We see that this functional form resembles the data of Rauch *et al.* (1975) reproduced in Fig. 1.7. Their results say that a neutron spin precession of 4π corresponds to a spatial integral of the magnetic field of 1.49×10^{-4} tesla meters. And this agrees with a measurement of Fig. 1.7 by a ruler. When we put numbers into Eq. (1.6.8), we find that θ in radians is indeed 4π,

$$\theta = \frac{3.82608(5.05078 \times 10^{-27})1.49 \times 10^{-4}}{1.05457 \times 10^{-34}(2.1736 \times 10^3)} = 12.56. \quad (1.6.13)$$

Thus, the rough calculations presented in this Ramble are in agreement with the spin-$\frac{1}{2}$ neutron's ket requiring a phase change of 4π for the phase to return to its initial value.

Basdevant and Dalibard (2006, Chap. 3) delve into neutron interferometry and touch on the experiment of Werner *et al.* (1975).

Ramble 2 — The spin precession of the muon

This Ramble describes an experiment by Sandweiss *et al.* (1973) that measures the spin precession of the spin-$\frac{1}{2}$ positive muon. The following discussion is not self-contained, but the cited references and Townsend (2000, Sec. 4.3) should fill the gaps. The measurement was done as part of an experiment on muon polarization and time-reversal invariance. Additional material on muon spin rotation is found in the books by Schenck (1985) and Yaouanc and de Réotier (2011). Accelerator muon beams are used to probe matter and a vast literature on applications is available.

The positive muons, μ^+, resulted from the decay of K_L^0 mesons. These mesons were generated in the zero-gradient synchrotron at the Argonne National Laboratory. The K_L^0 meson is the longer-lived linear combination of the K^0 and $\bar{K}^0$ mesons (Ho-Kim and Pham, 1998, Sec. 11.1). The K_L^0 meson decays according to

$$K_L^0 \rightarrow \pi^- + \mu^+ + \nu_\mu. \tag{1.6.14}$$

Here π^- is a negative pi-meson and ν_μ is a muon neutrino. The muon loses energy as it moves within an aluminum cylinder and eventually comes to rest within it.

The muon then finds itself in a constant magnetic field of about 0.006 tesla and the spin vector of the muon precesses about the direction of this constant magnetic field. The muon has a lifetime of 2.2 μs and decays into a positron, a neutrino and an anti-neutrino,

$$\mu^+ \rightarrow e^+ + \nu_e + \bar{\nu}_\mu. \tag{1.6.15}$$

The positron is preferentially emitted in the direction of the muon's spin vector. Thus, the detection of the positrons as a function of time reveals the projection of the muon spin vector onto the plane perpendicular to the magnetic field $\bar{B}$.

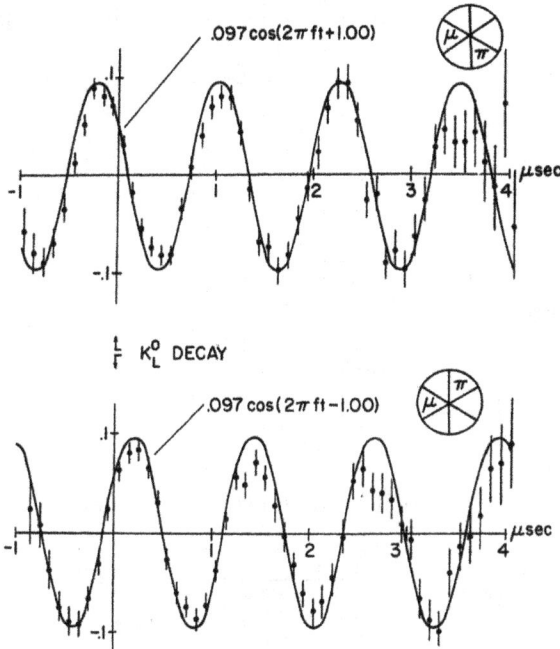

Figure 1.8. The muon spin precession data of Sandweiss *et al.* (1973) from two of their detectors. The precession frequency is 807.5 kHz. Reprinted figure with permission from Sandweiss *et al.* (1973). Copyright 1973 by the American Physical Society.

Figure 1.8 presents examples of the experimental data from Sandweiss *et al.* (1973) for two of their detectors. The predicted oscillatory behavior of Eqs. (1.3.33) and (1.3.34) is evident. Sandweiss and colleagues find the frequency f of the oscillations is 0.8075 MHz for Fig. 1.8. They use a functional time dependence of

$$0.097 \cos(2\pi f t \pm 1.00), \tag{1.6.16}$$

which adds a phase shift to Eq. (1.3.33). The frequency of the fit leads to an angular frequency of

$$\Omega = 5.0737 \times 10^6 \text{ radians/s.} \tag{1.6.17}$$

A careful use of a ruler with Fig. 1.8 leads to 0.80 MHz, which is quite close to the quoted 0.8075 MHz.

We next compare the magnetic field with what is expected from

$$\Omega = g_\mu \mu_\mu |\bar{B}|/\hbar, \tag{1.6.18}$$

with $g_\mu = 2.00233$,

$$\mu_\mu = \frac{e\hbar}{2m_\mu}, \tag{1.6.19}$$

and the subscript μ indicates quantities for the muon. We again use App. 1 to find

$$\mu_\mu = \frac{1.602176 \times 10^{-19}(1.05457 \times 10^{-34})}{2(1.88353 \times 10^{-28})} \text{ J/T} = 4.485 \times 10^{-26} \text{ J/T}. \tag{1.6.20}$$

We solve Eq. (1.6.18) for the magnetic field to get

$$|\bar{B}| = \frac{5.0737 \times 10^6(1.05457 \times 10^{-34})}{2.00233(4.485 \times 10^{-26})} \text{ T} = 0.596 \times 10^{-2} \text{ T}, \tag{1.6.21}$$

which agrees with the 0.006 tesla quoted by Sandweiss *et al.* (1973). Thus, we conclude that the muon spin vector precesses in agreement with the theory of Sec. 1.3.

References

G. Badurek, H. Rauch, A. Zeilinger, W. Bauspiess, and U. Bonse, "Phase-shift and spin-rotation phenomena in neutron interferometry", *Phys. Rev. D* **14**, 1177–1181 (1976).

R. Baierlein, *Thermal Physics* (Cambridge University Press, Cambridge, 1999), reprinted 2000.

J.-L. Basdevant and J. Dalibard, *The Quantum Mechanics Solver: How to Apply Quantum Theory to Modern Physics*, 2nd ed. (Springer-Verlag, Berlin, 2006).

B. H. Bransden and C. J. Joachain, *Quantum Mechanics*, 2nd ed. (Prentice Hall, Harlow, England, 2000).

C. Cohen-Tannoudji, B. Diu, and F. Laloë, *Quantum Mechanics*, Vols. 1 and 2 (John Wiley and Sons, New York, 1977).

A. Einstein, "Zur Quantentheorie de Strahlung", *Phy. Z.* **18**, 121 (1917).

Q. Ho-Kim and X.-Y. Pham, *Elementary Particles and Their Interactions: Concepts and Phenomena* (Springer-Verlag, Berlin, 1998).

E. Klempt, "Observability of the sign of the wave function", *Phys. Rev. D* **13**, 3125–3129 (1976).

K. Konishi and G. Paffuti, *Quantum Mechanics: A New Introduction* (Oxford University Press, Oxford, 2009).

F. Mezei, ed., *Neutron Spin Echo Spectroscopy: Basics, Trends and Applications* (Springer-Verlag, New York, 2010).

H. Rauch, A. Zeilinger, G. Badurek, A. Wilfing, W. Bauspiess, and U. Bonse, "Verification of coherent spinor rotation of fermions", *Physics Letters* **54A**, 425–427 (1975).

A. Rigamonti and P. Carretta, *Structure of Matter: An Introductory Course with Problems and Solutions*, 2nd ed. (Springer-Verlag Italia, Milan, 2009).

R. W. Robinett, *Quantum Mechanics: Classical Results, Modern Systems, and Visualized Examples*, 2nd ed. (Oxford University Press, Oxford, 2006).

J. J. Sakurai and J. Napolitano, *Modern Quantum Mechanics*, 2nd ed. (Addison-Wesley, Boston, 2011).

J. Sandweiss, J. Sunderland, W. Turner, W. Willis, and L. Keller, "Muon polarization in the decay $K_L^0 \to \pi^- \mu^+ \nu_\mu$, an experimental test of time-reversal invariance", *Phys. Rev. Lett.* **30**, 1002–1006 (1973).

A. Schenck, *Muon Spin Rotation Spectroscopy: Principles and Applications in Solid State Physics* (Adam Hilger Ltd., Bristol, 1985).

N. Tombros, C. Jozsa, M. Popinciuc, H. T. Jonkman, and B. J. van Wees, "Electronic spin transport and spin precession in single graphene layers at room temperature", *Nature* **448**, 571–574 (2007).

J. S. Townsend, *A Modern Approach to Quantum Mechanics* (University Science Books, Sausalito, CA, 2000).

S. A. Werner, R. Colella, A. W. Overhauser, and C. F. Eagen, "Observation of the phase shift of a neutron due to precession in a magnetic field", *Phys. Rev. Lett.* **35**, 1053–1055 (1975).

D. T. Wilkinson and H. R. Crane, "Precision measurement of the g factor of the free electron", *Phys. Rev.* **130**, 852–863 (1963).

J. Wilson and J. F. B. Hawkes, *Optoelectronics: An Introduction*, 2nd ed. (Prentice Hall, New York, 1989).

A. Yaouanc and P. Dalmas de Réotier, *Muon Spin Rotation, Relaxation, and Resonance: Applications to Condensed Matter* (Oxford University Press, Oxford, 2011).

Chapter 2

Two-Level Systems with a
Time-Independent Interaction

Chapter 1 shows how spin precession occurs for a spin-$\frac{1}{2}$ particle in a constant magnetic field. This leads to time-dependent expectation values of the components of the spin vector. More generally, time dependence in Quantum Mechanics is concerned with transitions between the states, or energy levels, of a system such as an atom, or a molecule. The Time-Dependent Schrödinger Equation (TDSE) provides a description of these transitions. But its solution is usually numerical and this limits insight into how the system develops over time. When we are able to restrict our considerations to only two energy levels, we can reduce the TDSE to two coupled first-order differential equations in time. And these we can solve! The two states may correspond to the spin up and spin down states of a spin-$\frac{1}{2}$ particle. Or the states may correspond to the energy levels of a double-well potential. Such two-state systems are quite useful in understanding magnetic resonance phenomena and the energy levels for molecules with inversion spectra such as ammonia. In addition, numerous settings arise where two energy levels provide a good approximation to a system with many energy levels. Such systems include atoms, molecules, nanostructures and even defects in solids.

In this and the following chapters, we treat systems with two energy levels. We first develop the case where the two states are coupled by a potential that is independent of time and the Sudden

Approximation is assumed valid. The TDSE is reduced to two coupled first-order ordinary differential equations in Sec. 2.1. The solutions are found and illustrated with plots of the oscillations of the populations of the two states in Secs. 2.2 and 2.3. Selected applications are then introduced in Sec. 2.4. These sections use the Sudden Approximation, which is discussed in Sec. 2.5 and App. 4. The chapter ends with Ramble 3 in Sec. 2.6, which is a brief introduction to systems that display random telegraph noise. Such systems have at least two energy levels and the transition from the lower to the higher energy level is often due to thermal excitations that act like an impulse. While such systems formally have time-dependent Hamiltonians, the fleeting nature of the impulse allows us to treat such systems in this chapter. Time-dependent coupling potentials provide a richer variety of phenomena and we start their study in Chap. 3.

Let the equations flow!

2.1. Formulation of the Two-Level Problem

We first find the eigenkets for the time-independent Hamiltonian $\hat{H}_0$ and then add a potential that couples the two states. We start with a two-level system and the Hamiltonian $\hat{H}_0$ with

$$\hat{H}_0|\phi_1\rangle = H_1|\phi_1\rangle, \qquad (2.1.1)$$

$$\hat{H}_0|\phi_2\rangle = H_2|\phi_2\rangle, \qquad (2.1.2)$$

with eigenkets $|\phi_i\rangle$ and their respective eigenvalues H_i. We take level 1 to be the ground state and level 2 to be the excited state. We assume that $|\phi_1\rangle$ and $|\phi_2\rangle$ form a complete set of states for our two-state system and

$$\langle\phi_i|\phi_j\rangle = \delta_{ij}, \qquad (2.1.3)$$

with δ_{ij} the Kronecker delta function. The TDSE is

$$i\hbar\frac{\partial|\psi_i(t)\rangle}{\partial t} = \hat{H}_0|\psi_i(t)\rangle, \qquad (2.1.4)$$

and $\hbar$ is Planck's constant divided by 2π with i equal to the square root of -1.

We define

$$|\psi_i(t)\rangle = a_i(t)|\phi_i\rangle, \tag{2.1.5}$$

and insert Eq. (2.1.5) into Eq. (2.1.4). We let $\hat{H}_0$ operate on $|\phi_i\rangle$ and d/dt on $a_i(t)$, and then multiply by $\langle\phi_i|$ from the left to find

$$i\hbar\frac{da_i(t)}{dt} = H_i a_i(t). \tag{2.1.6}$$

The solution to Eq. (2.1.6) is

$$a_i(t) = a_i e^{-iH_i t/\hbar}, \tag{2.1.7}$$

as may be directly verified. The most general state is written

$$|\psi(t)\rangle = a_1 e^{-iH_1 t/\hbar}|\phi_1\rangle + a_2 e^{-iH_2 t/\hbar}|\phi_2\rangle, \tag{2.1.8}$$

and the a_i are determined by the initial condition of the two-level system at $t = 0$.

We now introduce a potential $\hat{V}_c$ that couples the two states $|\phi_1\rangle$ and $|\phi_2\rangle$ and induces transitions between them. Figure 2.1 shows how the energy levels change when $\hat{V}_c$ is "suddenly" turned on. This Sudden Approximation is treated in detail in Bransden and Joachain (2000, Sec. 9.5) and is outlined below in Sec. 2.5 and App. 4. The

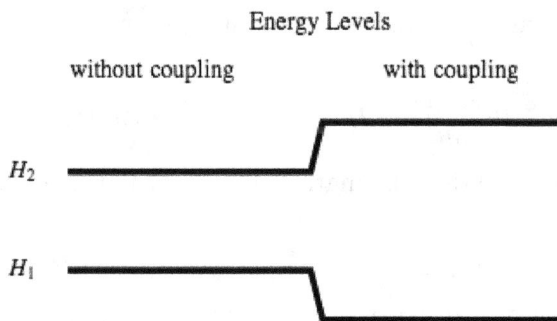

Figure 2.1. A schematic drawing of how the energy levels are before (*left*) and after (*right*) the coupling potential is activated.

Hamiltonian $\hat{H}$ is now

$$\hat{H} = \hat{H}_0 + \hat{V}_c, \tag{2.1.9}$$

and we need to solve

$$i\hbar \frac{\partial |\psi(t)\rangle}{\partial t} = \hat{H} |\psi(t)\rangle. \tag{2.1.10}$$

Since the $|\phi_i\rangle$ form a complete set of states for our two-level system, we generalize Eq. (2.1.8) and write

$$|\psi(t)\rangle = c_1(t)|\phi_1\rangle + c_2(t)|\phi_2\rangle. \tag{2.1.11}$$

We start to find the $c_i(t)$ by inserting Eq. (2.1.11) into Eq. (2.1.10),

$$i\hbar \frac{dc_1(t)}{dt}|\phi_1\rangle + i\hbar \frac{dc_2(t)}{dt}|\phi_2\rangle = c_1(t)H_1|\phi_1\rangle + c_1(t)\hat{V}_c|\phi_1\rangle$$

$$+ c_2(t)H_2|\phi_2\rangle + c_2(t)\hat{V}_c|\phi_2\rangle. \tag{2.1.12}$$

Next we assume that

$$\langle \phi_i|\hat{V}_c|\phi_i\rangle = 0. \tag{2.1.13}$$

In many cases, these matrix elements vanish due to a symmetry such as parity or we may redefine the H_i to include the value of the matrix element. Here, the former is assumed. We first multiply Eq. (2.1.12) from the left with $\langle \phi_1|$ and use Eq. (2.1.3) to find

$$i\hbar \frac{dc_1(t)}{dt} = H_1 c_1(t) + \langle \phi_1|\hat{V}_c|\phi_2\rangle c_2(t). \tag{2.1.14}$$

Then we return to Eq. (2.1.12) and multiply from the left with $\langle \phi_2|$ to find

$$i\hbar \frac{dc_2(t)}{dt} = H_2 c_2(t) + \langle \phi_2|\hat{V}_c|\phi_1\rangle c_1(t). \tag{2.1.15}$$

Let us also assume that the matrix element of $\hat{V}_c$ is real and negative. We write

$$V = \langle \phi_1|\hat{V}_c|\phi_2\rangle = \langle \phi_2|\hat{V}_c|\phi_1\rangle, \tag{2.1.16}$$

and we recast Eqs. (2.1.14) and (2.1.15) in matrix form

$$i\hbar \frac{d}{dt}\begin{pmatrix} c_1(t) \\ c_2(t) \end{pmatrix} = \begin{pmatrix} H_1 & V \\ V & H_2 \end{pmatrix}\begin{pmatrix} c_1(t) \\ c_2(t) \end{pmatrix}. \tag{2.1.17}$$

Finally, we utilize our key assumption that V is independent of time! This leads us to try

$$c_i(t) = c_i e^{-iEt/\hbar}, \tag{2.1.18}$$

where we need to find the values of E.

When our trial functions are substituted into Eq. (2.1.17), the $e^{-iEt/\hbar}$ appears on both sides and is canceled. We are left with an eigenvalue problem,

$$\begin{pmatrix} H_1 & V \\ V & H_2 \end{pmatrix} \begin{pmatrix} c_1 \\ c_2 \end{pmatrix} = E \begin{pmatrix} c_1 \\ c_2 \end{pmatrix}, \tag{2.1.19}$$

to solve. The determinant is

$$(H_1 - E)(H_2 - E) - V^2 = 0, \tag{2.1.20}$$

which is

$$E^2 - E(H_1 + H_2) + H_1 H_2 - V^2 = 0. \tag{2.1.21}$$

This quadratic equation in E yields

$$E = (H_1 + H_2)/2 \pm \{[(H_1 - H_2)/2]^2 + V^2\}^{1/2}. \tag{2.1.22}$$

We denote the two solutions as E_+ and E_-. The square root contains the dependence on the coupling potential's matrix element V. We define the average

$$E_a = (H_1 + H_2)/2, \tag{2.1.23}$$

and the difference

$$\Delta = (H_2 - H_1)/2. \tag{2.1.24}$$

So

$$(E_\pm - E_a)/\Delta = \pm\{1 + [V/\Delta]^2\}^{1/2}, \tag{2.1.25}$$

provides the plot of Fig. 2.2 to show how the V-dependence splits the two energy levels. The plot uses the magnitude V of the coupling matrix element and we see the energy levels diverge as the magnitude of V increases.

Section 2.2 develops the eigenkets of Eq. (2.1.10) for the present case of Eq. (2.1.19).

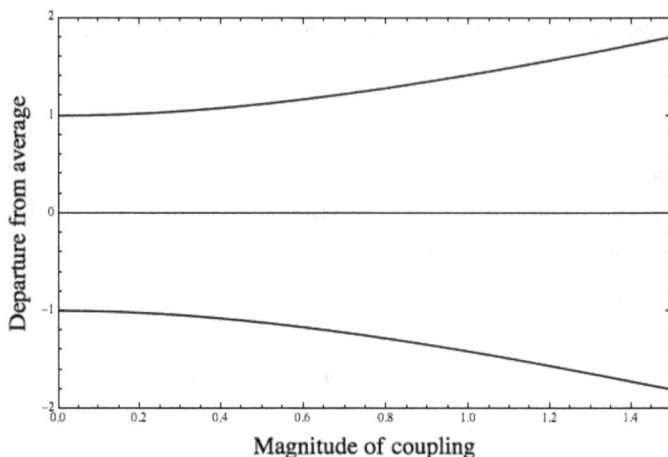

Figure 2.2.   The departure of the energy levels from the average given by Eq. (2.1.23) versus the magnitude of the coupling potential. The value of Δ is assumed to be 1.0 here, with $H_2 = +1$ and $H_1 = -1$.

2.2. Finding the Eigenkets

The next task is to construct the eigenkets that correspond to E_+ and E_-, that is, find the coefficients c_1 and c_2 of Eq. (2.1.18). These will allow us to determine the probability the system is in a given state as a function of time. Please recall that the coupling potential matrix element V between states 1 and 2 is independent of time. Several paths are possible. The first is the straightforward approach that puts $E_\pm$ into the matrix equation, Eq. (2.1.19). One then enters a sea of algebra as shown in App. 3. We take a second path that is based on Secs. 4.2 and 4.6 of Bittner (2010). This permits simpler expressions for the probabilities. Our notation and signs differ from Bittner.

   E_a and Δ are defined in Eqs. (2.1.23) and (2.1.24), respectively, and we assume Δ is greater than or equal to zero. Then

$$H_1 = \frac{H_1}{2} + \frac{H_2}{2} - \frac{H_2}{2} + \frac{H_1}{2} = E_a - \Delta, \qquad (2.2.1)$$

$$H_2 = \frac{H_2}{2} + \frac{H_1}{2} - \frac{H_1}{2} + \frac{H_2}{2} = E_a + \Delta. \qquad (2.2.2)$$

These definitions allow us to rewrite Eq. (2.1.19) with the 2×2 identity matrix I as

$$E_a I \begin{pmatrix} c_1 \\ c_2 \end{pmatrix} + \begin{pmatrix} -\Delta & V \\ V & \Delta \end{pmatrix} \begin{pmatrix} c_1 \\ c_2 \end{pmatrix} = E \begin{pmatrix} c_1 \\ c_2 \end{pmatrix}, \qquad (2.2.3)$$

where E is the eigenenergy $E_\pm$, which we now write in the compact form

$$E_\pm = E_a \pm (\Delta^2 + V^2)^{1/2}. \qquad (2.2.4)$$

We want the time-independent eigenkets $|\psi_\pm\rangle$ corresponding to $E_\pm$, respectively, to be orthonormal,

$$\langle \psi_+ | \psi_- \rangle = 0, \qquad (2.2.5)$$

$$\langle \psi_\pm | \psi_\pm \rangle = 1. \qquad (2.2.6)$$

The $|\psi_\pm\rangle$ are built with the $|\phi_i\rangle$ parallel to Eq. (2.1.11) and we recall the $|\phi_i\rangle$ are orthonormal. Thus, to satisfy Eqs. (2.2.5) and (2.2.6), we require, respectively,

$$c_{1+}c_{1-} + c_{2+}c_{2-} = 0, \qquad (2.2.7)$$

$$(c_{1\pm})^2 + (c_{2\pm})^2 = 1. \qquad (2.2.8)$$

We note with all the elements of Eq. (2.2.3) real, we take the c_i to be real.

We return to Eq. (2.2.3) and start the solution by subtracting the E_a term from both sides. This leaves

$$\begin{pmatrix} -\Delta & V \\ V & +\Delta \end{pmatrix} \begin{pmatrix} c_1 \\ c_2 \end{pmatrix} = \pm \sqrt{\Delta^2 + V^2} \begin{pmatrix} c_1 \\ c_2 \end{pmatrix}. \qquad (2.2.9)$$

We need to solve for $c_{1\pm}$ and $c_{2\pm}$. The plus sign yields the coupled equations

$$[-\Delta - \sqrt{\Delta^2 + V^2}]c_{1+} + V c_{2+} = 0, \qquad (2.2.10)$$

$$V c_{1+} + [\Delta - \sqrt{\Delta^2 + V^2}]c_{2+} = 0. \qquad (2.2.11)$$

Equation (2.2.10) gives

$$c_{1+} = V c_{2+}/[\Delta + \sqrt{\Delta^2 + V^2}]. \tag{2.2.12}$$

When this is substituted into Eq. (2.2.8), we find

$$c_{2+}^2 = (\Delta + \sqrt{\Delta^2 + V^2})^2[V^2 + (\Delta + \sqrt{\Delta^2 + V^2})^2]^{-1}. \tag{2.2.13}$$

Hence,

$$c_{1+} = V/[V^2 + (\Delta + \sqrt{\Delta^2 + V^2})^2]^{1/2}, \tag{2.2.14}$$

$$c_{2+} = (\Delta + \sqrt{\Delta^2 + V^2})/[V^2 + (\Delta + \sqrt{\Delta^2 + V^2})^2]^{1/2}. \tag{2.2.15}$$

These have the appearance of a sine and a cosine, but which is which? It helps to let the matrix element V go to zero. Then we see $E_+ \to H_2$, $c_{1+} \to 0$ and $c_{2+} \to 1$. Thus, we set

$$c_{1+} = \sin\theta, \tag{2.2.16}$$

$$c_{2+} = \cos\theta. \tag{2.2.17}$$

These lead to a surprising result for $\tan 2\theta$. First, we evaluate

$$\sin 2\theta = 2\sin\theta\cos\theta$$
$$= 2V(\Delta + \sqrt{\Delta^2 + V^2})/[V^2 + (\Delta + \sqrt{\Delta^2 + V^2})^2], \tag{2.2.18}$$

and

$$\cos 2\theta = \cos^2\theta - \sin^2\theta$$
$$= 2\Delta(\Delta + \sqrt{\Delta^2 + V^2})/[V^2 + (\Delta + \sqrt{\Delta^2 + V^2})^2]. \tag{2.2.19}$$

These yield

$$\tan 2\theta = V/\Delta. \tag{2.2.20}$$

This result proves useful when the occupation probabilities of the states are sought.

Similarly, the minus sign in Eq. (2.2.9) produces

$$(-\Delta + \sqrt{\Delta^2 + V^2})c_{1-} + Vc_{2-} = 0, \tag{2.2.21}$$

$$Vc_{1-} + (\Delta + \sqrt{\Delta^2 + V^2})c_{2-} = 0. \tag{2.2.22}$$

The last equation is

$$c_{1-} = -(\Delta + \sqrt{\Delta^2 + V^2})c_{2-}/V, \tag{2.2.23}$$

and with Eq. (2.2.8)

$$c_{2-}^2 = V^2/[V^2 + (\Delta + \sqrt{\Delta^2 + V^2})^2]. \tag{2.2.24}$$

Equation (2.2.7) is satisfied if c_{2-} is taken with the negative square root, then the same angle θ arises and

$$c_{1-} = \cos\theta, \tag{2.2.25}$$

$$c_{2-} = -\sin\theta. \tag{2.2.26}$$

We note that as the matrix element V goes to zero, $E_- \rightarrow H_1$, $c_{1-} \rightarrow 1$ and $c_{2-} \rightarrow 0$. These results are consistent with those for E_+.

We use the $c_{i\pm}$ to construct a matrix M that connects the $|\psi_\pm\rangle$ and the $|\phi_i\rangle$. M follows from Eqs. (2.2.25–26) and (2.2.16–17),

$$\begin{pmatrix} |\psi_-\rangle \\ |\psi_+\rangle \end{pmatrix} = \begin{pmatrix} \cos\theta & -\sin\theta \\ \sin\theta & \cos\theta \end{pmatrix} \begin{pmatrix} |\phi_1\rangle \\ |\phi_2\rangle \end{pmatrix}. \tag{2.2.27}$$

We note the inverse matrix is

$$M^{-1} = \begin{pmatrix} \cos\theta & \sin\theta \\ -\sin\theta & \cos\theta \end{pmatrix}, \tag{2.2.28}$$

and matrix multiplication verifies

$$MM^{-1} = \begin{pmatrix} 1 & 0 \\ 0 & 1 \end{pmatrix}. \tag{2.2.29}$$

Appendix 3 shows that M does, indeed, diagonalize the matrix in Eq. (2.2.9).

Now we are ready to calculate the occupation probabilities in Sec. 2.3.

2.3. The Occupation Probabilities

The eigenkets have been found in Sec. 2.2 for the two-level system with a time-independent coupling potential. We now examine how the wave function for the system evolves in time. We use Eq. (2.1.18) to recall our definitions

$$c_{i\pm}(t) = e^{-iE_{\pm}t/\hbar}c_{i\pm}. \tag{2.3.1}$$

The state of the two-level system with the matrix element $V \neq 0$ is

$$
\begin{aligned}
|\psi(t)\rangle &= c_+(t)|\psi_+\rangle + c_-(t)|\psi_-\rangle \\
&= e^{-iE_+t/\hbar}c_+|\psi_+\rangle + e^{-iE_-t/\hbar}c_-|\psi_-\rangle. \tag{2.3.2}
\end{aligned}
$$

We next need to expose the $|\phi_i\rangle$ that lie buried in Eq. (2.3.2).

Let us start the system at $t = 0$ in level 1. We assume the coupling potential $\hat{V}_c$ is then switched on instantly. This is another example of the Sudden Approximation (Bransden and Joachain, 2000, Sec. 9.5). The initial condition is found with the aid of

$$\begin{pmatrix} |\phi_1\rangle \\ |\phi_2\rangle \end{pmatrix} = M^{-1} \begin{pmatrix} |\psi_-\rangle \\ |\psi_+\rangle \end{pmatrix}, \tag{2.3.3}$$

to be

$$|\psi(0)\rangle = |\phi_1\rangle = \cos\theta|\psi_-\rangle + \sin\theta|\psi_+\rangle. \tag{2.3.4}$$

This state ket is a superposition of the eigenkets of $\hat{H}_0 + \hat{V}_c$. We next use Eq. (2.2.28) and, for t > 0, we apply Eq. (2.3.2) to find

$$|\psi(t)\rangle = e^{-iE_-t/\hbar}\cos\theta|\psi_-\rangle + e^{-iE_+t/\hbar}\sin\theta|\psi_+\rangle. \tag{2.3.5}$$

We make the $|\phi_i\rangle$ explicit with Eq. (2.2.27)

$$
\begin{aligned}
|\psi(t)\rangle &= e^{-iE_-t/\hbar}\cos\theta(\cos\theta|\phi_1\rangle - \sin\theta|\phi_2\rangle) \\
&\quad + e^{-iE_+t/\hbar}\sin\theta(\sin\theta|\phi_1\rangle + \cos\theta|\phi_2\rangle). \tag{2.3.6}
\end{aligned}
$$

We assume a measurement of the system at time t will tell us if the system is in level 1 or in level 2. So, we regroup Eq. (2.3.6)

$$
\begin{aligned}
|\psi(t)\rangle &= (e^{-iE_-t/\hbar}\cos^2\theta + e^{-iE_+t/\hbar}\sin^2\theta)|\phi_1\rangle \\
&\quad + \cos\theta\sin\theta(-e^{-iE_-t/\hbar} + e^{-iE_+t/\hbar})|\phi_2\rangle. \tag{2.3.7}
\end{aligned}
$$

We first multiply from the left by $\langle\phi_2|$ to find

$$\langle\phi_2|\psi(t)\rangle = -i\sin 2\theta \sin(\sqrt{\Delta^2 + V^2}t/\hbar)e^{-iE_a t/\hbar}. \qquad (2.3.8)$$

Here we inserted the values of E_+ and E_- from Eq. (2.1.22) and used the definitions of E_a and Δ from Eqs. (2.1.23) and (2.1.24), respectively.

Thus, the probability of the measurement finding the system in level 2 is

$$p_2(t) = |\langle\phi_2|\psi(t)\rangle|^2 = \sin^2 2\theta \sin^2(\sqrt{\Delta^2 + V^2}t/\hbar). \qquad (2.3.9)$$

Equation (2.2.20) for $\tan 2\theta$ provides a simple expression for $\sin 2\theta$,

$$\sin 2\theta = V/\sqrt{\Delta^2 + V^2}, \qquad (2.3.10)$$

and we write

$$p_2(t) = (V^2/(\Delta^2 + V^2))\sin^2 \omega_R t, \qquad (2.3.11)$$

with the angular Rabi frequency ω_R defined as

$$\omega_R = \sqrt{\Delta^2 + V^2}/\hbar. \qquad (2.3.12)$$

The Rabi frequency specifies how often the system changes from level 1 to level 2 and back. When $\Delta \neq 0$ and the system starts in level 1, then $p_2(t)$ starts from 0 and reaches a maximum of $V^2/(\Delta^2 + V^2)$ before decreasing. The Rabi frequency depends on the magnitude V of the matrix element of the coupling potential and the separation between the unperturbed energy levels. Further, Eq. (2.3.11) leads to

$$p_2(t) \propto t^2 \quad \text{for } t \to 0. \qquad (2.3.13)$$

We return to this point in Chap. 3.

In a similar fashion, we note that Eq. (2.3.7) leads to

$$\langle\phi_1|\psi(t)\rangle = e^{-iE_- t/\hbar}\cos^2 \theta + e^{-iE_+ t/\hbar}\sin^2 \theta, \qquad (2.3.14)$$

and this leads to

$$p_1(t) = |\langle\phi_1|\psi(t)\rangle|^2 = 1.0 - \sin^2 2\theta \sin^2(\sqrt{\Delta^2 + V^2}t/\hbar). \qquad (2.3.15)$$

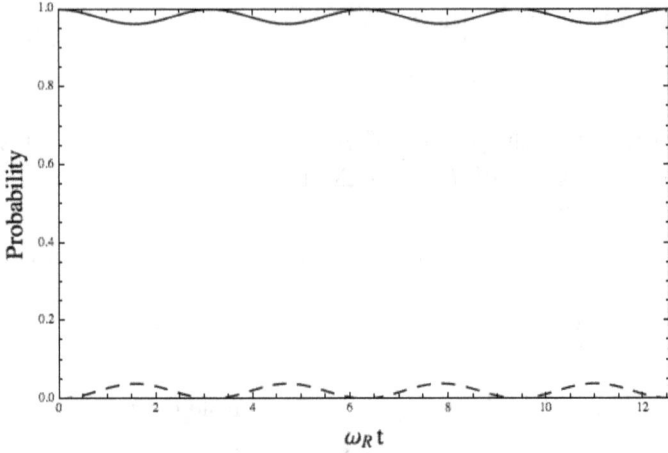

Figure 2.3. The occupation probabilities $p_1(t)$ (solid curve) and $p_2(t)$ (dashed curve) versus time for $V/\Delta = 0.2$ for a two-level system with a time-independent coupling potential.

And with Eq. (2.3.9), we see that

$$p_1(t) = 1.0 - p_2(t), \qquad (2.3.16)$$

which is reassuring.

Plots of $p_1(t)$ and $p_2(t)$ appear in Figs. 2.3 and 2.4 for V/Δ equal to 0.2 and 0.9, respectively. We remark that the system is in level 1 at $t = 0$. For small V/Δ, $p_1(t)$ is approximately 1.0 and $p_2(t)$ is approximately 0. This follows from the coefficient in front of the $\sin^2$ term of Eq. (2.3.11). The range of the variation of the $p_i(t)$ increases with an increase in V/Δ, and both reach an extremum of 0.5 when $V = \Delta$. Further increases in V/Δ lead to $p_1(t)$ and $p_2(t)$ approaching 0 and 1, respectively. Finally, for $\Delta = 0$ or $V/\Delta = \infty$, we have Fig. 2.5. Here, both $p_i(t)$ vary over the entire interval of 0 to 1.

We note that the above formalism requires the coupling potential $\hat{V}_c$ to be non-zero in order to have transitions between levels 1 and 2. If the system is in level 2 at time t and the coupling potential is "turned off", then the system remains in level 2. The system does not spontaneously relax to the lower level, level 1. The inclusion of spontaneous emission and other relaxation mechanisms awaits

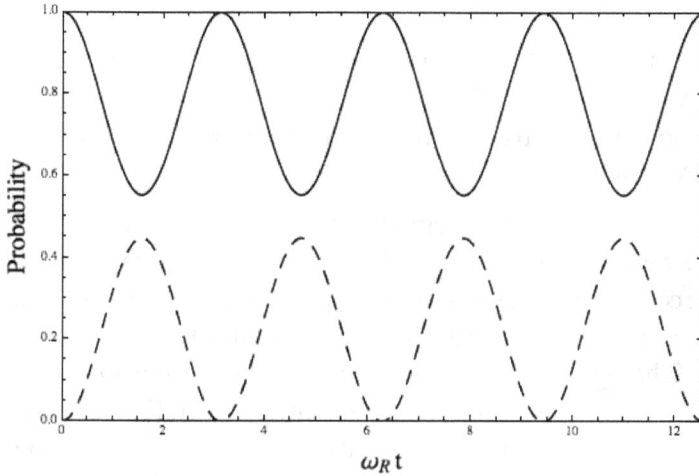

Figure 2.4. The occupation probabilities $p_1(t)$ (solid curve) and $p_2(t)$ (dashed curve) versus time for $V/\Delta = 0.9$ for a two-level system with a time-independent coupling potential.

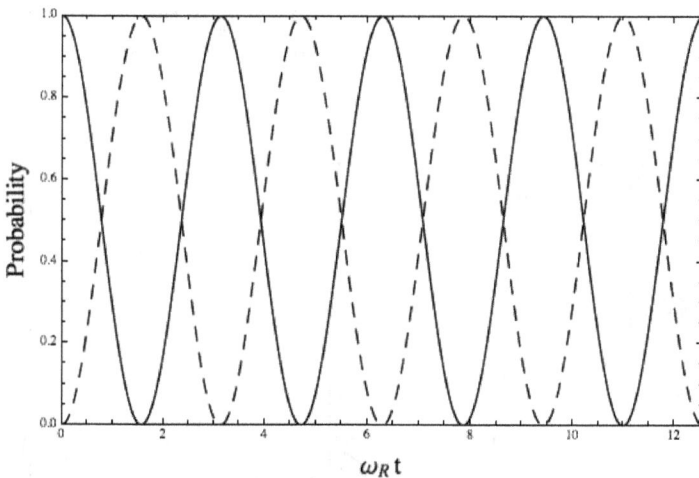

Figure 2.5. The occupation probabilities $p_1(t)$ (solid curve) and $p_2(t)$ (dashed curve) versus time for $V/\Delta = \infty$, which is $\Delta = 0$, for a two-level system with a time-independent coupling potential.

further developments in Chap. 5. The curious reader may consult a Quantum Optics text such as Berman and Malinovsky (2011).

2.4. Examples with a Time-Independent Coupling Potential

Several molecules are described by the formalism developed in this chapter for two-level systems. For example, ammonia, which is NH_3, has three hydrogen atoms in an equilateral triangle. The nitrogen atom is either above or below the plane defined by the three hydrogen atoms. These two configurations are equivalent and we may picture this situation as a double-well potential as in Fig. 2.6. Here the coordinate z represents the location of the nitrogen atom with respect to the x-y plane defined by the three hydrogen atoms. The equilibrium locations for the nitrogen atom are $z_0 = \pm 0.038$ nm and the potential barrier height V_0 at $z = 0$ is approximately 257 meV. These data are taken from Bransden and Joachain (2000, pp. 737–739). When the potential barrier is high and/or wide, the energy levels of the two potential wells are identical and independent. However, for a potential barrier similar to that pictured in Fig. 2.6, the nitrogen atom tunnels between the potential wells (Townsend, 2000, Sec. 4.5).

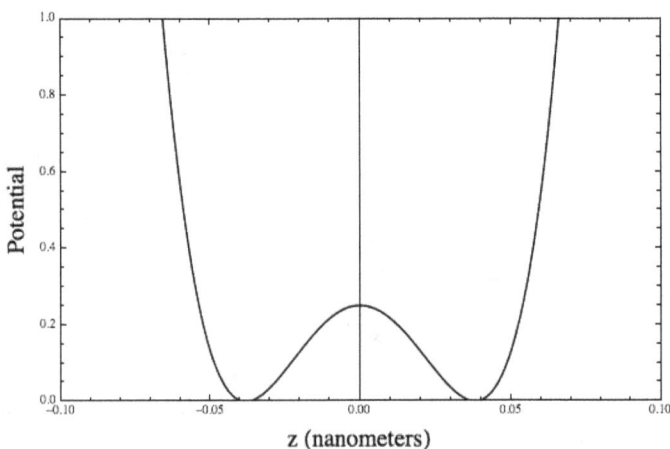

Figure 2.6. An approximate double-well potential in one dimension for the N atom in NH_3.

We use the earlier results of this chapter to see how tunneling splits the low-lying energy levels of the double-well potential. Tunneling is represented by a time-independent coupling potential. We employ Eqs. (2.1.19) to (2.1.22) and we treat the lowest energy level. Here H_1 and H_2 are the unperturbed energy levels of the left and right potential wells, respectively. Now $H_1 = H_2$ and we set $V = -V'$ with $V' > 0$. The latter makes the sign explicit. Equation (2.1.22) leads us to the energy levels with the coupling due to quantum mechanical tunneling,

$$E_{\pm} = H_1 \pm V'. \tag{2.4.1}$$

Thus, the unperturbed ground state is now split into two levels that form an inversion doublet.

The ammonia molecule undergoes an oscillation between the energy levels corresponding to $H_1 - V'$ and $H_1 + V'$. This means the ammonia molecule emits and absorbs radiation with an energy equal to $2V'$. The probability of the nitrogen atom being in the left or right potential well follows from Eqs. (2.3.11) and (2.3.15). The nitrogen atom is assumed to start in the left well at time 0 and we use the subscripts L and R in place of 1 and 2 in the kets. Since $H_1 = H_2$ here, $\Delta = 0$, and the equations simplify to

$$p_L(t) = \cos^2(V't/\hbar), \tag{2.4.2}$$

$$p_R(t) = \sin^2(V't/\hbar). \tag{2.4.3}$$

We digress to discuss the wave functions before continuing with the radiation aspects of ammonia. The sine and cosine of θ in Eqs. (2.2.25) and (2.2.26) become

$$\cos\theta = 1/\sqrt{2}, \tag{2.4.4}$$

$$\sin\theta = -1/\sqrt{2}. \tag{2.4.5}$$

This leads to the following version of Eq. (2.2.27)

$$\begin{pmatrix} |\psi_-\rangle \\ |\psi_+\rangle \end{pmatrix} = \begin{pmatrix} 1/\sqrt{2} & 1/\sqrt{2} \\ -1/\sqrt{2} & 1/\sqrt{2} \end{pmatrix} \begin{pmatrix} |\phi_L\rangle \\ |\phi_R\rangle \end{pmatrix}$$

$$= (1/\sqrt{2}) \begin{pmatrix} |\phi_L\rangle + |\phi_R\rangle \\ -|\phi_L\rangle + |\phi_R\rangle \end{pmatrix}. \tag{2.4.6}$$

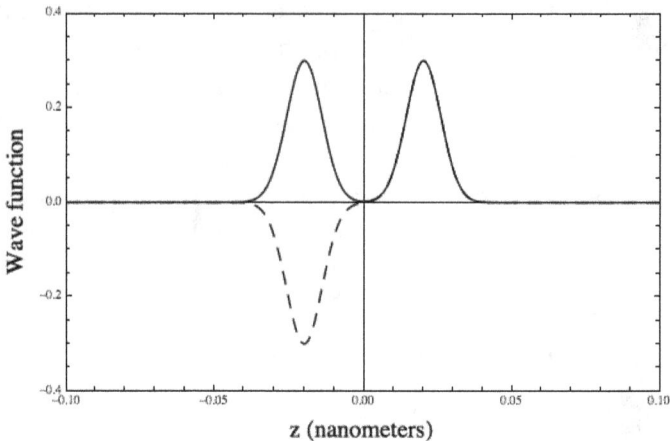

Figure 2.7. Approximate wave functions for the doublet states. Symmetric state (solid curve) and anti-symmetric state (dashed curve). Both states coincide for positive z. These illustrative wave functions are not normalized.

The state $|\psi_-\rangle$ is symmetric upon the exchange of L and R and has the energy $H_1 - V'$, so it is the lower energy state. The state $|\psi_+\rangle$ is anti-symmetric and its energy is $H_1 + V'$, so it is higher in energy. The wave functions plotted in Fig. 2.7 are based on Gaussians.

We return to the energy levels of ammonia. The splitting of the two energy levels is $2V'$ and this is found to be 9.84×10^{-2} meV or 0.793 cm^{-1} (Rigamonti and Carretta, 2009, pp. 303–304). This corresponds to a frequency ν of 23.8 gigahertz and a wavelength λ equal to 1.26×10^{-2} m or 1.26 cm, which is in the microwave part of the electromagnetic spectrum. The time to go from $p_L = 1.$ to $p_L = 0.$ is estimated with the aid of Eq. (2.4.2) by setting

$$V't/\hbar = \pi/2, \qquad (2.4.7)$$

or

$$t = h/(4V') \approx 1.05 \times 10^{-11} \text{ s.} \qquad (2.4.8)$$

It is essential here to put V' into joules or h into eV-s.

This inversion doublet transition of ammonia was observed by Cleeton and Williams (1934) in work that initiated the field of microwave spectroscopy. They used absorption, which is caused by

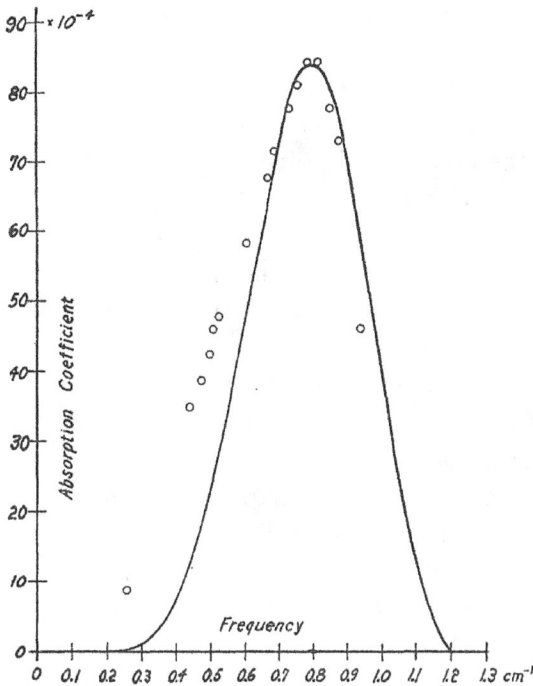

Figure 2.8. The absorption curve for ammonia from Cleeton and Williams (1934). The solid line is a theoretical curve described in their paper and the open circles are their measured data. The vertical axis goes from 0 to 90×10^{-4}/cm and the horizontal axis covers 0 to 1.3 cm^{-1}.

transitions from the lower energy level to the higher energy level of the doublet. We observe that the above λ of 1.26 cm corresponds to 0.793 cm^{-1} and this falls within the peak region in Cleeton and Williams (1934, Fig. 4), which is reproduced here as Fig. 2.8. Astronomers have used this 23.8 GHz transition to identify ammonia in the interstellar medium (Cheung *et al.*, 1968).

We note that the other low-lying levels of ammonia are also split by tunneling, as are those of similar molecules. The energy levels above the ground state "see" a reduced potential energy barrier, so that the amount of splitting increases. This is illustrated in Bransden and Joachain (2000, Fig. 16.9) and Rigamonti and Carretta (2009, p. 303). The latter also remark that deuterated ammonia, ND$_3$, has a reduced splitting of 6.3×10^{-3} eV and the resulting frequency is

1.59 GHz. Thus, the magnitude of the splitting is dependent on the potential energy barrier height in Fig. 2.6, the width of this potential barrier and the masses of the atoms. This topic is explored further in Hecht (2000, Chap. 37), where the WKB approximation to tunneling is applied to the double-well potential for ammonia.

We continue with ammonia and add an external electric field, since the resultant energy levels are still described by the equations of Chap. 2. The ammonia molecule has an electric dipole moment μ_e due to the nitrogen atom attracting electrons, thereby leaving the hydrogen atoms slightly positive. This transition dipole moment for the lowest-lying inversion doublet is 1.471 debyes (Takami *et al.*, 1979 and references therein). Now, 1 debye = 3.336×10^{-30} coulomb-meters, so this μ_e converts to 4.91×10^{-30} coulomb-meters. We return to this ammonia transition in Chap. 4 when masers are considered. But here we find the energies of the two levels due to the coupling of the dipole moment to an electric field $\bar{E}$. The matrix elements of $\hat{\mu}_e \cdot \bar{E}$ are evaluated in the $\{|\phi_1\rangle, |\phi_2\rangle\}$ basis. Each is only non-zero in either the left or right potential well. Thus, the off-diagonal matrix elements of $\hat{\mu}_e \cdot \bar{E}$ are zero and the diagonal matrix elements of $\hat{\mu}_e \cdot \bar{E}$ may be non-zero. (We note that when a basis such as $\{|\psi_+\rangle, |\psi_-\rangle\}$ is used, the non-zero matrix elements are off-diagonal. This is shown in Townsend (2000, Sec. 13.5).)

This coupling adds the energy $\pm\mu_e|\bar{E}|$ to the diagonal elements of Eqs. (2.1.17) and (2.1.19). The sign of the dipole moment μ_e changes when the nitrogen atom crosses the plane of the three hydrogen atoms. We assume that $|\mu_e\bar{E}| \ll V'$ and this requires an electric field magnitude less than 10^5 V/m or so. This electric field leads to $|\mu_e\bar{E}| = 3.06 \times 10^{-6}$ eV as compared to the doublet splitting of 9.84×10^{-5} eV.

The doublet energy levels come from the determinant

$$\begin{vmatrix} H_1 + \mu_e|\bar{E}| - E & -V' \\ -V' & H_1 - \mu_e|\bar{E}| - E \end{vmatrix} = 0, \qquad (2.4.9)$$

or

$$E_\pm = H_1 \pm \sqrt{\mu_e^2|\bar{E}|^2 + V'^2}. \qquad (2.4.10)$$

With our assumption that $|\mu_e \bar{E}| \ll V'$, we expand the square root to find

$$E_\pm = H_1 \pm V' \pm \mu_e^2 |\bar{E}|^2 / 2V'. \qquad (2.4.11)$$

When an electric field is pointed in the z direction, it exerts a force F_z on an ammonia molecule

$$F_z = -\frac{\partial E_\pm}{\partial z} = -\left(\pm \mu_e^2 \frac{|\bar{E}|}{V'} \frac{\partial |\bar{E}|}{\partial z} \right). \qquad (2.4.12)$$

Thus, if the electric field varies in the z direction, ammonia molecules traversing the region with the electric field will be separated into two beams, one beam with the molecules in the E_+ state and the second beam with the molecules in the E_- state. This is the electric dipole moment equivalent of the Stern–Gerlach experiment (Townsend, 2000, Chap. 1) for magnetic dipole moments and it allows us to sort ammonia molecules. We see in Chap. 4 that this separation is quite useful for initiating the ammonia maser.

We next delve into the Sudden Approximation, which we have used repeatedly in this chapter.

2.5. The Sudden Approximation

It is common to encounter situations in Quantum Mechanics where the interactions and/or the coupling potential change suddenly. Admittedly, this instantaneous change is an idealization, but it has led to the Sudden Approximation used in this chapter. This approximation is thoroughly discussed in Bransden and Joachain (2000, Sec. 9.5), who also treat the case where the transition is of a finite duration in time. This lets them explain when the Sudden Approximation is appropriate. To a rough approximation, the Sudden Approximation is reasonable when the switching time is much less than $\hbar$, Planck's constant/2π, divided by a relevant energy difference. Duffey (1992, Chap. 8) presents an alternative argument based on the time evolution operator and the present App. 4 extends his approach and finds a similar condition results. If we assume an energy difference of one electron-volt, then the switching time needs to be much less than 6.6×10^{-16} s. Even with an energy difference of 10^{-4} eV, a switching time of 6.6×10^{-12} s is still an experimental challenge for most systems. But the Sudden Approximation continues to be used because it allows significant simplifications and often yields reasonable results. We apply it here to a two-level system that becomes a three-level system when the depth of the potential well is increased. This calculation prepares us for Ramble 3 on random telegraph noise or signals in Sec. 2.6.

Let the Hamiltonian for $t < 0$ be

$$\hat{H}_< = \hat{H}_0 + \hat{V}_< = \hat{H}_0 - V_<, \tag{2.5.1}$$

with $V_<$ constant and positive. The similar Hamiltonian for $t > 0$ is

$$\hat{H}_> = \hat{H}_0 + \hat{V}_> = \hat{H}_0 - V_>. \tag{2.5.2}$$

Here $\hat{H}_0$ is the kinetic energy operator. The eigenkets are written in a generalization of Eq. (2.1.8) as

$$|\psi_<(t)\rangle = \sum_{i=1}^{m} a_i e^{-iH_<^{(i)}t/\hbar}|\theta_i\rangle, \tag{2.5.3}$$

and

$$|\psi_>(t)\rangle = \sum_{j=1}^{n} b_j e^{-iH_>^{(j)}t/\hbar}|\phi_j\rangle, \tag{2.5.4}$$

with the energy eigenvalues in the arguments of the exponentials. Equations (2.5.3) and (2.5.4) go with the Hamiltonians in Eqs. (2.5.1) and (2.5.2), respectively. Yes, the notation is a challenge. And we are neglecting the continuum states of these Hamiltonians, i.e., the states for positive energy. Our kets are based on the bound states of the above Hamiltonians and are orthonormal,

$$\langle\theta_k|\theta_l\rangle = \delta_{kl}, \tag{2.5.5}$$

and

$$\langle\phi_k|\phi_l\rangle = \delta_{kl}. \tag{2.5.6}$$

The Sudden Approximation consists of equating the eigenkets at $t = 0$, the time for the change from $\hat{H}_<$ to $\hat{H}_>$. This leads us to

$$\sum_{i=1}^{m} a_i|\theta_i\rangle = \sum_{j=1}^{n} b_j|\phi_j\rangle, \tag{2.5.7}$$

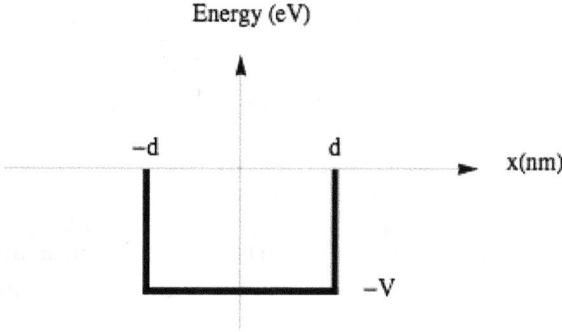

Figure 2.9. One-dimensional square-well potential for the Sudden Approximation calculations. Space is divided into 3 regions: $1 = -\infty$ to $-d$, $2 = -d$ to d
and $3 = d$ to ∞. Here $d = 0.3$ nm.

and we are able to determine the probability of the system being in a
specific state $|\phi_l\rangle$ for $t > 0$ through finding b_l. We multiply Eq. (2.5.7)
from the left by $\langle\phi_l|$, use Eq. (2.5.6), and see

$$\sum_{i=1}^{m} a_i\langle\phi_l|\theta_i\rangle = \sum_{j=1}^{n} b_j\langle\phi_l|\phi_j\rangle = b_l. \tag{2.5.8}$$

We note that overlap integrals are required.

We illustrate the Sudden Approximation with a finite square-well
in one spatial dimension that runs from $-d$ to d and is pictured in
Fig. 2.9 with $d = 0.3$ nm. The potential well depth is varied at $t = 0$
and is selected so that there are only two energy levels for $t < 0$. Then
the potential well is deepened to have three energy levels for $t > 0$. We
now need to solve the time-independent Schrödinger Equation to find
the wave function in each of the three regions indicated in Fig. 2.9,

$$-\frac{\hbar^2}{2m}\frac{d^2\varphi}{dx^2} - V\varphi = -|E|\varphi, \tag{2.5.9}$$

or

$$\frac{d^2\varphi}{dx^2} + \frac{2m}{\hbar^2}(V - |E|)\varphi = 0, \tag{2.5.10}$$

and the electron mass is used. For $x < -d$, with $V = 0$ and $\kappa =
\sqrt{2m|E|}/\hbar$, we have

$$\varphi_1(x) = Ae^{+\kappa x}. \tag{2.5.11}$$

Within the potential well, we assume solutions either symmetric or anti-symmetric in x with $k = \sqrt{2m(V - |E|)}/\hbar$. The ground state and the second excited state are symmetric and the wave functions have the form

$$\varphi_2(x) = B \cos kx, \qquad (2.5.12)$$

while the first excited state is anti-symmetric in x and its wave function is proportional to $\sin kx$. For $x > d$, we find with $V = 0$,

$$\varphi_3(x) = De^{-\kappa x}. \qquad (2.5.13)$$

All the coefficients and the energy are different for each excited state. We use a superscript prime to distinguish the excited states from the ground state.

We are led to the eigenvalue equations when we match the wave functions and their derivatives at the edges of the potential well. For the symmetric $\varphi_2(x)$, we have

$$k \tan kd = \kappa, \qquad (2.5.14)$$

and for the anti-symmetric $\varphi_2(x)$,

$$\kappa' \tan k'd = -k'. \qquad (2.5.15)$$

These lead to $\{\kappa, k\}$ for the ground state and $\{\kappa', k'\}$ for the first excited state, respectively.

We first work with the ground state. Then, by symmetry,

$$A = D. \qquad (2.5.16)$$

When we match at $x = d$, we find

$$B = De^{-\kappa d}/\cos kd. \qquad (2.5.17)$$

These results are useful for normalizing the wave function. We need three integrals and we note the wave function is real. The first integral is

$$I_1 = \int_{-\infty}^{-d} A^2 e^{2\kappa x} dx = \frac{A^2}{2\kappa} e^{-2\kappa d}, \qquad (2.5.18)$$

the second is

$$I_2 = \int_{-d}^{d} B^2 \cos^2(kx)dx$$

$$= \frac{B^2}{2} \int_{-d}^{d} (1 + \cos 2kx)dx$$

$$= B^2 \left(d + \frac{1}{2k} \sin 2kd \right), \tag{2.5.19}$$

and the third is

$$I_3 = \int_{d}^{\infty} D^2 e^{-2\kappa x} dx = \frac{D^2}{2\kappa} e^{-2\kappa d}. \tag{2.5.20}$$

We note that because of Eq. (2.5.16), we have $I_1 = I_3$.
    Now with

$$I_1 + I_2 + I_3 = 1, \tag{2.5.21}$$

we find

$$D^2 \frac{e^{-2\kappa d}}{\kappa} \left\{ 1 + \frac{\kappa}{\cos^2(kd)} \left(d + \frac{\sin 2kd}{2k} \right) \right\} = 1. \tag{2.5.22}$$

Thus, we have the wave function for the ground state. The first excited state introduces changes in the second integral,

$$I_2' = B'^2 \int_{-d}^{d} \sin^2(2k'x)dx = B'^2 \left(d - \frac{\sin 2k'd}{2k'} \right), \tag{2.5.23}$$

in addition to the introduction of primes in the other integrals, and the changed matching condition at $x = d$. Figure 2.10 shows the two wave functions for $V = 2.5$ eV and $t < 0$, while Fig. 2.11 displays the three wave functions for $V = 5.0$ eV and $t > 0$. The figures show that the ground state is symmetric in x and the first excited state is anti-symmetric in x. In addition, the second excited state of Fig. 2.11 is symmetric in x. It is apparent from these figures that the less tightly-bound states have wave functions with greater spatial extent.

    It is worth demonstrating that the ground state and the first excited state are orthogonal for each choice of V. We define these integrals as O_i for region i of Fig. 2.9 and we note the integral O_2

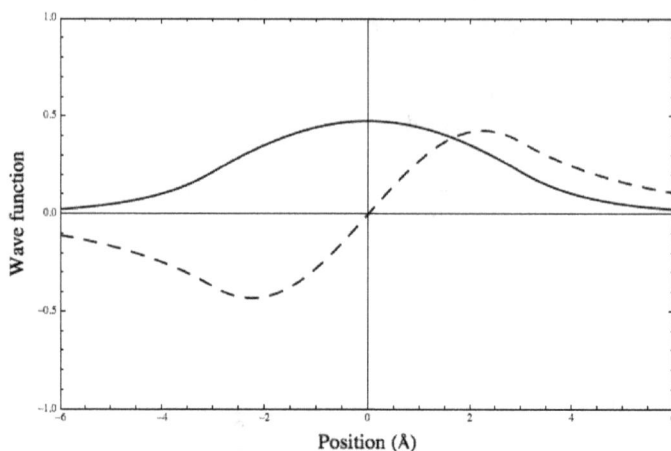

Figure 2.10. The normalized wave functions for the square-well potential with a depth of 2.5 eV: Ground state (solid curve) and excited state (dashed curve). The calculations were done with angstroms (Å), so the plotted wave functions are in units of $1/\text{Å}^{1/2}$. The total width of the square-well potential is 6 Å $= 0.6$ nm.

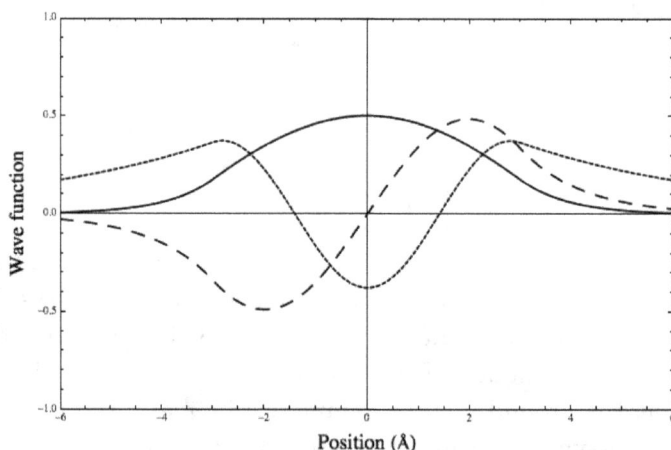

Figure 2.11. The normalized wave functions for the square-well potential with a depth of 5.0 eV: Ground state (solid curve), first excited state (long-dashed curve) and second excited state (short-dashed curve). The calculations were done with angstroms (Å), so the plotted wave functions are in units of $1/\text{Å}^{1/2}$. The total width of the square-well potential is 6 Å $= 0.6$ nm.

from $-d$ to d is the integral of a symmetric function times an anti-symmetric function, and this equals zero. We are left with

$$O_1 = \int_{-d}^{\infty} A e^{\kappa x} A' e^{\kappa' x} dx = \frac{AA'}{\kappa + \kappa'} e^{-(\kappa + \kappa')d}, \qquad (2.5.24)$$

and

$$O_3 = \int_{d}^{\infty} D e^{-\kappa x} D' e^{-\kappa' x} dx = \frac{DD'}{\kappa + \kappa'} e^{-(\kappa + \kappa')d}. \qquad (2.5.25)$$

We recall that $A = D$. The excited state is anti-symmetric and as Fig. 2.10 shows, we have $A' = -D'$ and $D' > 0$. This leads us to realize that $O_1 = -O_3$ and the ground state is orthogonal to the first excited state. A similar argument shows the first and second excited states for $V = 5.0$ eV are orthogonal. If we extend the integrals to cover the overlap of the ground state and the second excited state for $V = 5.0$ eV, we find they are orthogonal as expected.

We now rewrite Eq. (2.5.7) for our present two-level to three-level case,

$$a_1 |\theta_1\rangle + a_2 |\theta_2\rangle = b_1 |\phi_1\rangle + b_2 |\phi_2\rangle + b_3 |\phi_3\rangle. \qquad (2.5.26)$$

This is the $t = 0$ matching condition for the Sudden Approximation. The subscript 1 refers to the ground state and the subscripts 2 and 3 are for the excited states. Let us suppose we start in the ground state, so

$$a_1 = 1.0, \qquad (2.5.27)$$

$$a_2 = 0.0. \qquad (2.5.28)$$

We have constructed the wave functions above, so we are able to do the overlap integrals to get the b_i for the transitions into the energy levels for $t > 0$. We multiply from the left by $\langle \phi_i |$ and find

$$b_1 = \langle \phi_1 | \theta_1 \rangle, \qquad (2.5.29)$$

$$b_2 = \langle \phi_2 | \theta_1 \rangle = 0, \qquad (2.5.30)$$

and

$$b_3 = \langle \phi_3 | \theta_1 \rangle. \qquad (2.5.31)$$

The result for b_2 follows from the symmetry of the initial ground state and the anti-symmetry of the first excited state of the $V = 5.0$ eV potential well.

We break up the remaining overlap integrals into three pieces, T_i, each with one integral for each region shown in Fig. 2.9. We use a superscript double prime for the $t > 0$ state and find

$$T_1 = A'' A \int_{-\infty}^{-d} e^{(\kappa'' + \kappa)x} dx = A'' A \frac{e^{-(\kappa'' + \kappa)d}}{\kappa'' + \kappa}, \qquad (2.5.32)$$

$$T_2 = B'' B \int_{-d}^{d} \cos k'' x \cos kx dx$$

$$= B'' B \left\{ \frac{\sin(k'' - k)d}{k'' - k} + \frac{\sin(k'' + k)d}{k'' + k} \right\}, \qquad (2.5.33)$$

and

$$T_3 = D'' D \int_{d}^{\infty} e^{-(\kappa'' + \kappa)x} dx = A'' A \frac{e^{-(\kappa'' + \kappa)d}}{\kappa'' + \kappa}. \qquad (2.5.34)$$

We have used Eq. (2.5.16) in the last integral and we remark these expressions hold for both b_1 and b_3. Now we need numbers!

Table 2.1 has the energy levels for our two finite square-well cases. These come from numerically solving Eqs. (2.5.14) and (2.5.15).

We have enough information to evaluate the T_i numerically and/or in closed form, once we have the coefficients $\{A, B, D, A'', B'', D''\}$. When this is done we find the closed forms lead to

$$b_1 = 0.99448, \qquad (2.5.35)$$

$$b_3 = 0.0963953. \qquad (2.5.36)$$

The same numbers result when the numerical integrals are computed from $x = -1.0$ nm to $x = 1.0$ nm, that is, infinity is replaced by 1.0.

The transition probabilities follow from the squares of the b_i. The result for the ground state to ground state yields 0.98899, while the

Table 2.1. Energy levels for a finite square-well of 0.6 nm total width.

Potential depth (eV)	Ground state (eV)	First (eV)	Second (eV)
2.5	−1.987136	−0.63506	—
5.0	−4.379473	−2.60715	−0.22904

ground state to the second excited state provides 0.009292. These sum to 0.99828, which is less than 1.00. The discrepancy does not appear to be due to numerical errors. At the start of this section we stated we are ignoring the continuum states. It is time to reconsider. The t < 0 states are constructed according to Eqs. (2.5.27) and (2.5.28), so that

$$\langle \theta_1 | \theta_1 \rangle = 1.00, \quad (2.5.37)$$

in accord with Eq. (2.5.5). Let us insert the set of bound states $|\phi_i\rangle$ and assume they are a complete set. Then, we expect

$$\langle \theta_1 | \theta_1 \rangle = 1.00 = \langle \theta_1 | \sum_{k=1}^{3} |\phi_k\rangle \langle \phi_k | \theta_1 \rangle. \quad (2.5.38)$$

But, based on the above results, we find

$$\langle \theta_1 | \sum_{k=1}^{3} |\phi_k\rangle \langle \phi_k | \theta_1 \rangle = \langle \theta_1 | \phi_1 \rangle \langle \phi_1 | \theta_1 \rangle + \langle \theta_1 | \phi_3 \rangle \langle \phi_3 | \theta_1 \rangle = 0.99828.$$

$$(2.5.39)$$

We are led to conclude there is a slight probability of a transition from the ground state for $t < 0$ to a continuum state for $t > 0$. For completeness, we should consider the continuum states. However, for the present case, the bound states do reasonably well by themselves.

This section has introduced the Sudden Approximation and provided the details for a finite square-well example. This development is used in Sec. 2.6, Ramble 3, with random telegraph signals.

2.6. Ramble 3 on Random Telegraph Signals

The following chapters investigate time-dependent coupling potentials that are periodic in time. Between these and the time-independent cases of this chapter, there is a class of phenomena known as random telegraph signals (RTS) or random telegraph noise (RTN). Shore (1984) treats the modeling of such noise processes and has extensive references to the mathematics that is useful. Instead, we consider a simple model for RTS that involves the Sudden Approximation of Sec. 2.5. But first, we do a rapid survey of RTS

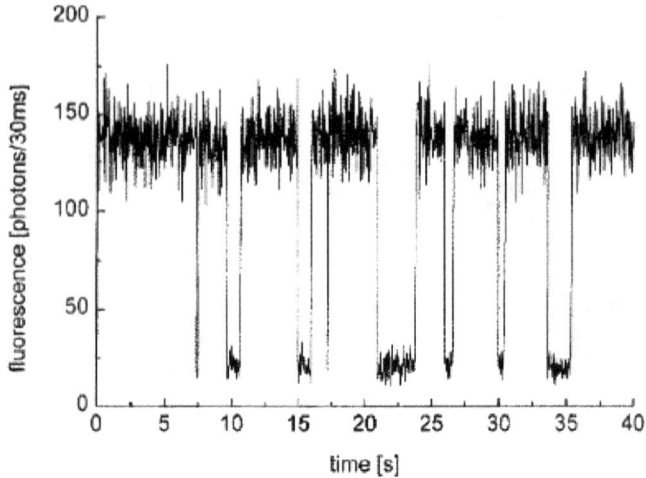

Figure 2.12. A fluorescence RTS from a calcium ion with a 30 ms integration time from Ritter and Eichmann (1997). Figure © IOP Publishing. Reproduced with permission. All rights reserved.

that starts with quantum jumps, a popular and useful type of RTS that provides data on electronic transitions in atoms or ions. We end with the 21 cm hydrogen emission line.

The last few decades have seen the development of the ability to observe the electromagnetic radiation emitted by an individual atom or molecule. Generally, random telegraph signals are seen as shown in Fig. 2.12 (Ritter and Eichmann, 1997, Fig. 4). We call the high level of the fluorescence signal the on-state and the low level the off-state. In this case, an ionized calcium atom, Ca^+, emits the fluorescence when an electron makes a transition from the $4^2P_{1/2}$ excited state to the $4^2S_{1/2}$ ground state. A laser tuned to 397 nm is used to excite the Ca^+ ion. So, during the on-state, the electron alternately absorbs and emits a photon. The off-states seen in Fig. 2.12 occur when the electron gets into the $3^2D_{5/2}$ state, whose transition to the ground state provides the quantum jump. Although the laser continues to illuminate the Ca^+ ion during the off-state, the laser does not induce transitions back to the ground state. Eventually, the Ca^+ ion relaxes to the ground state, the absorption of the laser radiation starts again, and we are back in the on-state.

Let us examine the on-state of Fig. 2.12 more closely. The $4^2P_{1/2}$ excited state is alternately occupied and empty, so the on-state is actually composed of a two-level RTS. The randomness enters in the time the Ca^+ ion remains in its ground state and in the time the excited state remains occupied. To discern this RTS, the photon detector needs to work on a faster scale than that used. The coupling potentials of these quantum jump experiments are time-dependent and are treated in Chaps. 3 to 5 for two levels.

We next consider two-level systems that show random telegraph signals and are found in condensed matter, in particular, devices made in silicon. The transition from the lower-energy state to the higher-energy state is generally thermally-activated or due to tunneling. The transitions occur on a time scale much faster then the time the system remains in the on-state or the off-state. Two examples of such RTS are current flow in a solid-state transistor and a time-dependent point defect in a pixel of a solid-state visible image sensor such as a charge-coupled device (CCD) (Janesick, 2001; Theuwissen, 1995). The image sensor data are obtained after an integration of the dark current for a specified time interval. The dark current is the current collected in a pixel in the absence of incident light.

We start with current flow in field-effect transistors (FET) (Taur and Ning, 1998) and random telegraph signals. The current flows in the channel of the FET, which is the part of the semiconductor adjacent to or at the semiconductor/dielectric interface. Now suppose there is a defect in the channel of the transistor or just adjacent to the channel in the gate dielectric. If such a defect is charged, it has the ability to affect the current flow in the FET. Successful characterization of individual defects became common in the 1980s when the devices shrank until each FET contained at most a few defects. Farmer *et al.* (1987), Uren *et al.* (1988) and Kandiah *et al.* (1989) contain examples. For simplicity, we assume the defect is either charged negative or neutral. If the defect is neutral, then the current is assumed to be in the high or on-state. But if the defect captures an electron, then the resulting negative charge of the defect repels electrons and the current decreases into the off-state. Other defect

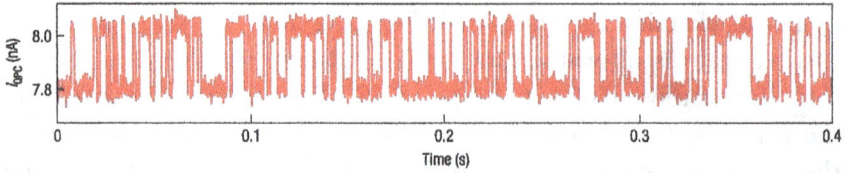

Figure 2.13. Current in nanoamperes versus time in seconds. The current in a quantum point contact is modulated by the charge on an adjacent quantum dot. The experimental details are in Sukhorukov *et al.* (2007, Fig. 1(c)). Reprinted by permission from Macmillan Publishers Ltd: Nature Physics, copyright 2007.

charge configurations are possible, but this defect with two states, a bistable defect, is the easiest to visualize. Actual mechanisms for the defects in FETs are discussed by, among others, Kandiah *et al.* (1989) and Andersson *et al.* (1992).

Figure 2.13 is a related example from Sukhorukov *et al.* (2007) wherein the random telegraph signals are found in a current that flows near a semiconductor quantum dot. An electron tunnels onto and off the quantum dot (Hanson, 2008) and the change in the charge on the quantum dot affects the nearby current flow. Here the quantum dot plays the role of a defect in a FET.

We now turn to RTS examples from CCDs and photodiodes (Streetman and Banerjee, 2006, p. 406). Prytherch (1996) presents early RTS data from a CCD pixel at $-6°C$. The responsible defect was not identified. Charged particle radiation damages CCDs and produces RTS as evidenced in Fig. 2.14 from Hopkins and Hopkinson (1995) and reported more recently by Tivarus and McColgin (2008). Neutrons induce defects in avalanche photodiodes and Buchinger *et al.* (1995) measured their dark current in the on-state and in the off-state. They related the differences in dark current to the presence of bistable defects, that is, defects that exist in either of two states. They found that the dark current differences are thermally-activated with an Arrhenius behavior. In addition, Buchinger and colleagues observed that the lifetimes of the two states had exponential distributions. They went on to deduce that the rates at which the defects switched from an on-state to an off-state and vice versa also show Arrhenius behaviors with activation energies, E, of 1.0 eV or

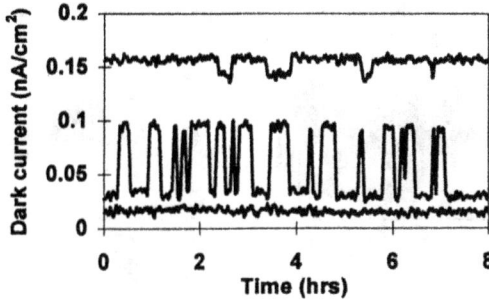

Figure 2.14. The dark current for 3 CCD pixels measured at 10°C after irradiation with 10 MeV protons (Hopkins and Hopkinson, 1995). The data for the bottom trace come from a pixel without RTS and with a low dark current level. The middle trace is from a pixel with a strong RTS. The top trace is the data from a high dark current pixel with an additional weak RTS. Figure reprinted with permission from the IEEE. Copyright 1995.

less. That is, the rate $R(T)$ is

$$R(T) = R_0 e^{-E/k_B T}, \qquad (2.6.1)$$

with a rate constant R_0, Boltzmann's constant k_B and the temperature T in kelvin. This applies to both their radiation-related defects and RTS defects present in their avalanche photodiodes that had not been irradiated with neutrons. The activation energy gives an idea of the energy barriers between the defect configurations for the on- and off-states. We note the dark current generation in CCDs is generally thermally-activated as shown by, for example, McColgin *et al.* (1995) and references therein.

Other distributions of on- and off-times have been observed. Stefani *et al.* (2009) point out that power-law distributions are often associated with nanoscale emitters. Kuno *et al.* (2001) study the fluorescence emission of ZnS-coated CdSe quantum dots, which are explained by Hanson (2008). The fluorescence is observed after a CdSe quantum dot is illuminated with the 488 nm line of an Ar$^+$ laser. Figure 2.15 is an example of the emission from a 2.7 nm quantum dot recorded by Kuno *et al.* (2001). The on and off periods are visible. The off times occur when the excited electron enters a non-radiative decay channel. In this particular case, the authors suggest tunneling into states associated with the surface of the

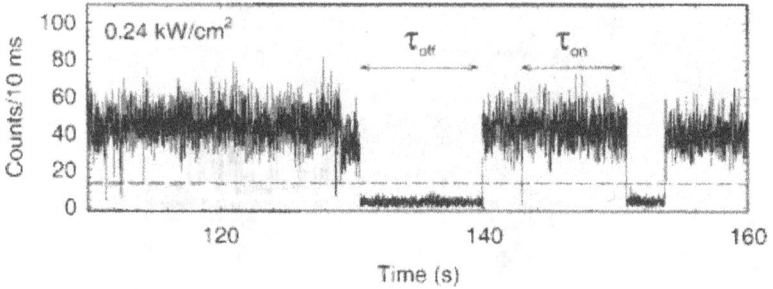

Figure 2.15. Fluorescence from a 2.7 nm quantum dot of CdSe with a ZnS overcoat (Kuno *et al.*, 2001, Fig. 3). The dashed line is the threshold for a fluorescence event. Counts are summed for 10 ms. High counts and low counts go with on- and off-times, respectively. Figure reprinted with permission of AIP Publishing LLC. Copyright 2001.

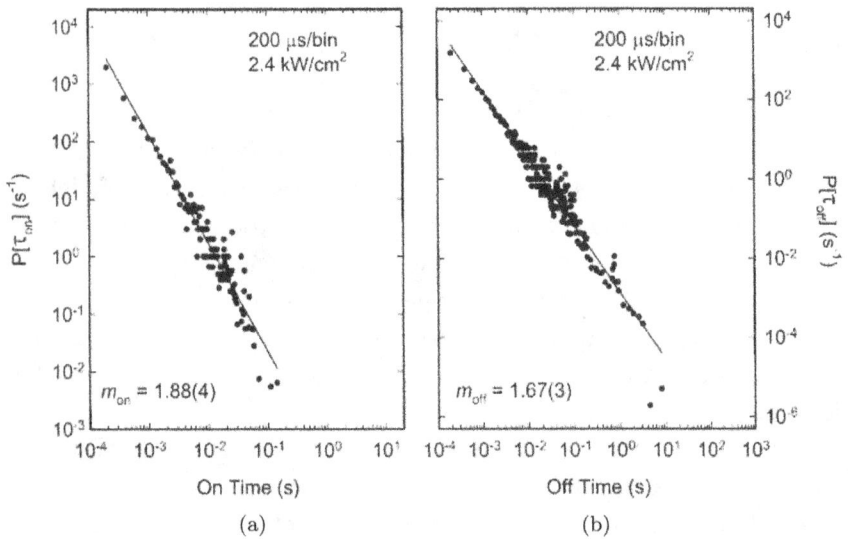

Figure 2.16. Distributions of on-times (a) and off-times (b) (Kuno *et al.*, 2001, Fig. 10). The power law distributions are seen on these log-log plots. Figure reprinted with permission of AIP Publishing LLC. Copyright 2001.

CdSe or the ZnS overcoat. Eventually, the electron returns to the quantum dot and the fluorescence resumes. Figure 2.16 presents the distributions of the on- and the off-times for a 2.7 nm quantum dot. Power laws are evident and are attributed to fluctuations in the environment of the quantum dot. These lead to a variation in the rates that the electron leaves and returns to the interior of the quantum dot proper.

Often the RTS examples appear to involve a bistable defect. The transition between the two states may involve, among other possibilities, the capture or emission of an electron, a change in the configuration of a defect, or a change in the atomic composition of a defect. An example of the last involves iron and boron in silicon and is B + Fe $\leftrightarrow$ FeB (Kimerling and Benton, 1983).

Interstitial iron is mobile in silicon around room temperature and is a strong generator of dark current at these temperatures (McColgin *et al.*, 1995), whereas FeB is a much weaker generator. This allowed the presence of interstitial iron to be tracked during anneals at 55°C in a CCD (McColgin *et al.*, 1997). When the iron paired with a boron atom, the dark current signal for the pixel dropped. And when the FeB dissociated, the dark current signal increased.

We may view these RTSs from CCDs and photodiodes as due to defects that generate dark current at two different rates. In semiconductors, dark current generation is often modeled by the excitation of a valence band electron into the conduction band. In silicon, this is usually treated as a two-step process. In the first step, a valence band electron is thermally excited to an energy level in the silicon band gap. The second step is the thermal excitation of this electron from the gap state into the conduction band. This process is described in texts on semiconductor devices, such as that of Grove (1967, Chap. 5), which provides a description of dark current generation and the related recombination events. We have two energy differences: the first is between the energy level in the gap and the valence band, and the second is between the conduction band and the energy level in the gap. Dark current generation is generally strongest when the two energy differences are similar in size. This motivates the following simple model for random telegraph noise.

We assume that in the on-state of the defect the energy differences are more similar than for the off-state. This leads to a higher dark current in the on-state. Let us consider a finite square-well as in Sec. 2.5. Hence, we are assuming the defect has a one-dimensional character. The change in the defect constitution or configuration is replaced by a change in the depth of the square-well and we make the further assumption that the Sudden Approximation is valid. We map the start of the continuum states of the square-well to the conduction band edge and the bound state of the square-well to the energy level in the silicon band gap. Thus, the off-state is represented by a square-well with an energy level deeper that half of the silicon band gap, which is about 0.56 eV at room temperature (Grove, 1967). We decrease the depth of the square-well to get an energy level closer to 0.56 eV. We emphasize that our model does not represent an actual defect in silicon, but rather shows the possible nature of a defect responsible for RTS. For example, the two potential wells may represent the two defect configurations discussed by Buchinger *et al.* (1995).

Table 2.2 has the energy levels for two finite square-well cases. These come from numerically solving Eqs. (2.5.14) and (2.5.15). We see that when the potential depth goes from 1.25 to 0.9 eV, the difference between the energy level and 0, our conduction band edge here, decreases. We identify the on-state with the potential depth of 0.9 eV and the off-state with the potential depth of 1.25 eV. This model is solely designed to give a crude idea of how such a change in the potential may explain the RTS.

We end this Ramble with a consideration of the hyperfine splitting of the 1s ground state of the hydrogen atom. The emission from the higher energy state to the lower energy state is at a wavelength of 21 cm and is used in Astrophysics to map the distribution of hydrogen in space. Murdin (2009, Chap. 52) has a short history and maps of hydrogen's distribution. The first observations of the 21 cm line were reported by Ewen and Purcell (1951) and Muller and Oort (1951).

Table 2.2.   Energy levels for a finite square-well of 0.6 nm total width.

Potential depth (eV)	Energy level (eV)
0.9	−0.555
1.25	−0.851

We introduce the Hamiltonian for the hyperfine splitting, find the energy separation and estimate the spontaneous emission lifetime of the higher energy state.

Cohen-Tannoudji *et al.* (1977, Chap. 12) shows that the Hamiltonian for hyperfine interactions has three terms. The first term describes the interaction of the nuclear magnetic moment of the proton with the electron and is proportional to the orbital angular momentum $\bar{L}$ of the electron. This term is zero for the 1s electron of hydrogen. The second term provides the interaction of the magnetic moments of the proton and the electron. The 1s orbital is spherically symmetric, so this term also contributes zero. The third term represents the interaction of the magnetic moment of the electron spin with the magnetic field inside a proton of finite size. This term is called Fermi's contact term and is

$$H_{hf} = -\frac{\mu_0}{4\pi}\left\{\frac{8\pi}{3}\hat{\bar{M}}_s \cdot \hat{\bar{M}}_I \delta(\bar{r})\right\}. \tag{2.6.2}$$

Here S is the electron's spin, I is the proton's spin, μ_0 is the vacuum permeability and the Dirac delta function operates on the spatial part of the wave function for the 1s electron (Sakurai and Napolitano, 2011, App. B),

$$\psi_{1s}(\bar{r}) = \frac{1}{\sqrt{4\pi}}\left(\frac{1}{a_0}\right)^{3/2} 2e^{-r/a_0}, \tag{2.6.3}$$

with a_0 the Bohr radius and here we use the reduced mass of the electron and the proton.

We also need the Dirac delta function for spherical coordinates

$$\delta(\bar{r}) = \frac{\delta(r)\delta(\theta)\delta(\phi)}{r^2 \sin\theta}. \tag{2.6.4}$$

The hyperfine states are labeled by the total angular momentum

$$\bar{F} = \bar{L} + \bar{S} + \bar{I} = \bar{S} + \bar{I}, \tag{2.6.5}$$

for the 1s state. Since $I = S = 1/2$, $F = 1$ or $F = 0$ and we have a triplet and a singlet, respectively. We will see that the singlet is the lower energy state. The energy of the hyperfine states is

$$\langle H_{hf} \rangle_{1s} = \frac{e^2 g_p}{3\varepsilon_0 c^2 m_e m_p} \langle \delta(\bar{r}) \rangle \langle F | \hat{\bar{I}} \cdot \hat{\bar{S}} | F \rangle$$

$$= \frac{e^2 g_p}{3\varepsilon_0 c^2 m_e m_p} \left(\frac{1}{\pi a_0^3} \right) \langle F | \hat{\bar{I}} \cdot \hat{\bar{S}} | F \rangle. \qquad (2.6.6)$$

Here e is the elementary charge, g_p is the proton's gyromagnetic ratio, ε_0 is the vacuum permittivity, c is the speed of light, m_e is the electron mass, m_p is the proton mass and $c^2 = 1/(\mu_0 \varepsilon_0)$ is used. The numerical values of these physical constants are defined in App. 1 along with $\hbar$, which is Planck's constant divided by 2π. The latter is used in the fine structure constant

$$\alpha = \frac{e^2}{4\pi\varepsilon_0 \hbar c}, \qquad (2.6.7)$$

that is used to reduce Eq. (2.6.6) to a number. The result with the reduced mass is

$$\langle H_{hf} \rangle_{1s} = 8.4780 \times 10^{43} \langle F | \hat{\bar{I}} \cdot \hat{\bar{S}} | F \rangle = A \langle F | \hat{\bar{I}} \cdot \hat{\bar{S}} | F \rangle. \qquad (2.6.8)$$

The next step is to evaluate the spin term with the aid of Eq. (2.6.5)

$$\bar{I} \cdot \bar{S} = \frac{1}{2} (\bar{F}^2 - \bar{I}^2 - \bar{S}^2) = \frac{1}{2} \{ F(F+1) - I(I+1) - S(S+1) \}. \qquad (2.6.9)$$

For the 1s state of hydrogen we find for $F = 1$,

$$\bar{I} \cdot \bar{S} = \frac{1}{2} \left(2 - \frac{3}{4} - \frac{3}{4} \right) \hbar^2 = \frac{1}{4} \hbar^2, \qquad (2.6.10)$$

and for $F = 0$,

$$\bar{I} \cdot \bar{S} = \frac{1}{2} \left(-\frac{3}{4} - \frac{3}{4} \right) \hbar^2 = -\frac{3}{4} \hbar^2. \qquad (2.6.11)$$

Thus, the energy difference between the $F = 1$ and the $F = 0$ hyperfine states is

$$\Delta_{hf} = \left(\frac{1}{4} - \left(-\frac{3}{4} \right) \right) A\hbar^2 = A\hbar^2 = 5.88 \times 10^{-6} \text{ eV}. \qquad (2.6.12)$$

We turn this into a frequency when we divide by $2\pi\hbar$ and find

$$\nu_{hf} = 1.4207 \times 10^9 \text{ Hz}, \qquad (2.6.13)$$

which leads to a wavelength of

$$\lambda_{hf} = 0.211 \text{ m} = 21.1 \text{ cm}. \qquad (2.6.14)$$

The experimental value found by Crampton *et al.* (1963) is

$$\nu_{hf} = 1420405751.800 \pm 0.028 \text{ Hz}. \qquad (2.6.15)$$

This amount of experimental accuracy is achieved with a hydrogen maser and the long lifetime of the $F = 1$ hyperfine state. Masers are discussed in Chap. 4 while we next estimate the lifetime.

The lifetime of the $|F = 1, 0\rangle$ state is the inverse of the Einstein coefficient A_{hf}, the coefficient for spontaneous emission, which we now estimate. This is definitely a side trip, but it is appropriate for a Ramble!

We use the relationships of Sec. 1.4 with energy level 1 the $|F = 0, 0\rangle$ state and energy level 2 the $|F = 1, 0\rangle$ state. In the absence of degeneracy, we have from Eq. (1.4.12)

$$B_{12} = B_{21}, \qquad (2.6.16)$$

and we next develop an approximation for the Einstein absorption coefficient B_{21}. Then by way of Eq. (1.4.13), we relate A_{hf} to $B_{hf} = B_{21}$ and we have

$$A_{hf} = \frac{\hbar\omega_{hf}^3}{\pi^2 c^3} B_{hf}, \qquad (2.6.17)$$

where $\omega_{hf} = 2\pi\nu_{hf}$.

We assume the electron in energy level 1 is excited to energy level 2 by magnetic dipole radiation. Then B_{hf} follows from the transition rate $W_{hf} = W_{21}$, which in turn comes from the amplitude for energy level 2, $c_2(t)$, in the expansion of the two-level wave function. The derivation parallels that for electric dipole radiation in Bransden and Joachain (2000, Secs. 11.2 and 11.3). First-order perturbation theory is used to get $c_2(t)$, which in turn leads to an approximation for the occupation probability $p_2(t)$ of the $|F = 1, 0\rangle$ state. The time derivative of $p_2(t)$ provides W_{hf} and the latter is multiplied by the speed of light and divided by the intensity distribution at the transition angular frequency ω_{hf} to get

$$B_{21}(\omega_{hf}) = B_{hf}(\omega_{hf}) = \frac{\pi \mu_0 \mu_e^2}{3\hbar^2}, \qquad (2.6.18)$$

with μ_e the Bohr magneton.

We put this result into Eq. (2.6.17) and find

$$A_{hf} = \frac{\hbar \omega_{hf}^3}{\pi^2 c^3} B_{hf} = \left(\frac{\hbar \omega_{hf}^3}{\pi^2 c^3} \right) \left(\frac{\pi}{3\hbar^2} \mu_0 \mu_e^2 \right) = 2.86 \times 10^{-15}/\text{s}.$$

$$(2.6.19)$$

It is apparent that the electron stays in the $F = 1$ hyperfine state around 0.35×10^{15} seconds! That is why this emission is included under random telegraph noise. In closing, we mention that the electron enters the $F = 1$ state due to collisions, absorption of a 21.1 cm photon, or as a result of the hydrogen atom undergoing decay from an excited state beyond the 1s states.

This completes our rapid survey of random telegraph signals. Chapter 3 takes us into the realm of time-dependent Hamiltonians.

References

G. I. Andersson, M. O. Andersson, and O. Engström, "Discrete conductance fluctuations in silicon emitter junctions due to defect clustering and evidence for structural changes by high-energy electron irradiation and annealing", *J. Appl. Phys.* **72**, 2680–2691 (1992).

P. R. Berman and V. S. Malinovsky, *Principles of Laser Spectroscopy and Quantum Optics* (Princeton University Press, Princeton, 2011).

E. R. Bittner, *Quantum Dynamics Applications in Biological and Materials Systems* (CRC Press, Boca Raton, 2010).

B. H. Bransden and C. J. Joachain, *Quantum Mechanics*, 2nd ed. (Prentice Hall, Harlow, England, 2000).

F. Buchinger, A. Kyle, J. K. P. Lee, C. Webb, and H. Dautet, "Identification of individual bistable defects in avalanche photodiodes", *Appl. Phys. Lett.* **66**, 2367–2369 (1995).

A. C. Cheung, D. M. Rank, C. H. Townes, D. D. Thornton, and W. J. Welch, "Detection of NH₃ molecules in the interstellar medium by their microwave emission", *Phys. Rev. Lett.* **21**, 1701–1705 (1968).

C. E. Cleeton and N. H. Williams, "Electromagnetic waves of 1.1 cm wave-length and the absorption spectrum of ammonia", *Phys. Rev.* **45**, 234–237 (1934).

C. Cohen-Tannoudji, B. Diu and F. Laloë, *Quantum Mechanics*, Vol. 2 (John Wiley and Sons, New York, 1977).

S. B. Crampton, D. Kleppner and N. F. Ramsey, "Hyperfine separation of ground-state atomic hydrogen", *Phys. Rev. Lett.* **11**, 338–340 (1963).

G. H. Duffey, *Quantum States and Processes* (Prentice Hall, Englewood Cliffs, N. J., 1992).

H. I. Ewen and E. M. Purcell, "Radiation from galactic hydrogen at 1,420 Mc/sec", *Nature* **168**, 356 (1951).

K. R. Farmer, C. T. Rogers, and R. A. Buhrman, "Localized-state interactions in metal-oxide-semiconductor tunnel diodes", *Phys. Rev. Lett.* **58**, 2255–2258 (1987).

A. S. Grove, *Physics and Technology of Semiconductor Devices* (John Wiley and Sons, New York, 1967).

G. W. Hanson, *Fundamentals of Nanoelectronics* (Pearson Prentice Hall, Upper Saddle River, N. J., 2008).

K. T. Hecht, *Quantum Mechanics* (Springer, New York, 2000).

I. H. Hopkins and G. R. Hopkinson, "Further measurements of random telegraph signals in proton irradiated CCDs", *IEEE Trans. Nucl. Sci.* **42**, 2074–2081 (1995).

J. R. Janesick, *Scientific Charge-Coupled Devices* (SPIE Press, Bellingham, Washington, 2001).

K. Kandiah, M. O. Deighton, and F. B. Whiting, "A physical model for random telegraph signal currents in semiconductor devices", *J. Appl. Phys.* **66**, 937–948 (1989).

L. C. Kimerling and J. L. Benton, "Electronically controlled reactions of interstitial iron in silicon", *Physica* **116B**, 297–300 (1983).

M. Kuno, D. P. Fromm, H. F. Hamann, A. Gallagher, and D. J. Nesbitt, "'On'/'off' fluorescence intermittency of single semiconductor quantum dots", *J. Chem. Phys.* **115**, 1028–1040 (2001).

W. C. McColgin, J. P. Lavine and C. V. Stancampiano, "Probing metal defects in CCD image sensors", *Mat. Res. Soc. Symp. Proc.* **378**, 713–724 (1995).

W. C. McColgin, J. P. Lavine, and C. V. Stancampiano, "Dark current spectroscopy of metals in silicon", *Mat. Res. Soc. Symp. Proc.* **442**, 187–192 (1997).

C. A. Muller and J. H. Oort, "The interstellar hydrogen line at 1,420 Mc/sec and an estimate of galactic rotation", *Nature* **168**, 357–358 (1951).

P. Murdin, *Secrets of the Universe: How We Discovered the Cosmos* (The University of Chicago Press, Chicago, 2009).

H. Prytherch, "Dark signal anomalies in the Kodak KAF 1300-L charge-coupled device", *Optical Engineering* **35**, 1796–1798 (1996).

A. Rigamonti and P. Carretta, *Structure of Matter: An Introductory Course with Problems and Solutions*, 2nd ed. (Springer-Verlag Italia, Milan, 2009).

G. Ritter and U. Eichmann, "Lifetime of the Ca^+ $3^2D_{5/2}$ level from quantum jump statistics of a single laser-cooled ion", *J. Phys. B: At. Mol. Opt. Phys.* **30**, L141–L146 (1997).

J. J. Sakurai and J. Napolitano, *Modern Quantum Mechanics*, 2nd ed. (Addison-Wesley, Boston, 2011).

B. W. Shore, "Modeling noise by jump processes in strong laser-atom interactions", *J. Optical Society of America B* **1**, 176–188 (1984).

F. D. Stefani, J. P. Hoogenboom, and E. Barkai, "Beyond quantum jumps: Blinking nanoscale light emitters", *Physics Today* **62**, February, 34–39 (2009).

B. G. Streetman and S. K. Banerjee, *Solid State Electronic Devices*, 6th ed. (Pearson Prentice Hall, Upper Saddle River, N. J., 2006).

E. V. Sukhorukov, A. N. Jordan, S. Gustavsson, R. Leturcq, T. Ihn and K. Ensslin, "Conditional statistics of electron transport in interacting nanoscale conductors", *Nature Phys.* **3**, 243–247 (2007).

M. Takami, H. Jones, and T. Oka, "Transition dipole moments of NH_3 in excited vibrational states determined by laser Stark spectroscopy", *J. Chem. Phys.* **70**, 3557–3558 (1979).

Y. Taur and T. H. Ning, *Fundamentals of Modern VLSI Devices* (Cambridge University Press, Cambridge, 1998).

A. J. P. Theuwissen, *Solid-State Imaging with Charge-Coupled Devices* (Kluwer Academic Publishers, Dordrecht, 1995).

C. Tivarus and W. C. McColgin, "Dark current spectroscopy of irradiated CCD image sensors", *IEEE Trans. Nucl. Sci.* **55**, 1719–1724 (2008).

J. S. Townsend, *A Modern Approach to Quantum Mechanics* (University Science Books, Sausalito, CA, 2000).

M. J. Uren, M. J. Kirton, and S. Collins, "Anomalous telegraph noise in small-area silicon metal-oxide-semiconductor field-effect transistors", *Phys. Rev. B* **37**, 8346–8350 (1988).

Chapter 3

Two-Level Systems with a Time-Dependent Interaction

Chapter 2 treated two-level systems with a time-independent coupling potential between the two energy levels. We now add time dependence to the coupling potential. This introduces a rich variety of phenomena when the time-dependent coupling potential represents the interaction of a classical electromagnetic field with our two-level system. For example, many atomic or molecular systems may be simplified to two energy levels if the interaction of other energy levels with the time-dependent electromagnetic field may be ignored. In addition, a static magnetic field splits an energy level into its magnetic sublevels. This leads to a two-level system for a spin-$\frac{1}{2}$ electron or nucleon, and this two-level system may be probed with a time-dependent electromagnetic field in search of electron spin or nuclear magnetic resonance, respectively.

We start in Sec. 3.1 with a cosine time dependence and turn the Time-Dependent Schrödinger Equation (TDSE) into a set of coupled ordinary differential equations. Further progress requires an approximation known as the Rotating Wave Approximation (RWA), which is developed in Sec. 3.2. The RWA permits closed-form expressions for the occupation probabilities of the two energy levels, p_1 and p_2, as functions of time. These are derived in Sec. 3.3 and compared to numerical solutions without the use of the RWA. The section ends with data showing an example of how $p_2(t)$ changes with the time t–Rabi oscillations. Then in Sec. 3.4, we return to the RWA and add a time-independent magnetic field. The final two sections present further closed-form solutions. Section 3.5 uses the RWA and the Interaction Picture. In Sec. 3.6 we investigate a circularly-polarized electromagnetic field, which has a time dependence that allows for closed-form solutions without the need for the RWA. Further examples of numerical solutions of the coupled equations appear in Shore (1990, Chaps. 3 and 4).

3.1. Developing the Coupled Equations

We assume the coupling potential operator $\hat{V}_c$ is proportional to the transition dipole moment $\hat{\bar{\mu}}_{21}$ between the two states represented by the kets $|\phi_1\rangle$ and $|\phi_2\rangle$. We take

$$\hat{V}_c = -\hat{\bar{\mu}}_{21} \cdot \bar{E} \cos \omega t, \tag{3.1.1}$$

with the time t and a time-dependent electromagnetic field of angular frequency ω. We also assume the wavelength of the electromagnetic field is large compared to the spatial extent of our system. This allows us to neglect spatial dimensions and is reasonable for systems of atomic size and smaller. We consider the electromagnetic field to be linearly polarized, so we work with the magnitude E and $\mu_{21} E$. The eigenenergies of the unperturbed Hamiltonian $\hat{H}_0$ are H_1 and H_2. We assume these are not altered by the electromagnetic field, although this point needs to be revisited when the electromagnetic field is strong (Cohen-Tannoudji *et al.*, 1992, Ch. VI). The complete Hamiltonian is

$$\hat{H} = \hat{H}_0 + \hat{V}_c. \tag{3.1.2}$$

With $\hbar$ = Planck's constant divided by 2π, we set

$$H_i = \hbar \omega_i, \tag{3.1.3}$$

and we work with the eigenkets

$$\hat{H}_0 |\phi_i\rangle = \hbar \omega_i |\phi_i\rangle. \tag{3.1.4}$$

In addition, we assume the $|\phi_i\rangle$ have definite parity and

$$\langle\phi_i|\phi_j\rangle = \delta_{ij}, \tag{3.1.5}$$

with δ_{ij} the Kronecker delta function.

We want to solve the TDSE

$$i\hbar\frac{\partial|\psi(t)\rangle}{\partial t} = \hat{H}|\psi(t)\rangle, \tag{3.1.6}$$

where $i = \sqrt{-1}$. Within our two-level system, we expand $|\psi(t)\rangle$

$$|\psi(t)\rangle = c_1(t)e^{-i\omega_1 t}|\phi_1\rangle + c_2(t)e^{-i\omega_2 t}|\phi_2\rangle, \tag{3.1.7}$$

since we find this form of time dependence to be useful. We assume

$$\langle\psi(t)|\psi(t)\rangle = 1, \tag{3.1.8}$$

which requires

$$|c_1(t)|^2 + |c_2(t)|^2 = 1, \tag{3.1.9}$$

in the light of Eq. (3.1.5).

We next substitute Eq. (3.1.7) into Eq. (3.1.6) and multiply from the left by $\langle\phi_i|$ as we did in Chap. 2, Sec. 1. This leads us to evaluate the diagonal matrix element of $\hat{V}_c$. The dipole moment operator $\hat{\bar{\mu}}$ is proportional to the spatial coordinate vector $\bar{r}$ or to a component of $\bar{r}$, so $\hat{\bar{\mu}}$ has odd parity. Then, by our assumption that the $|\phi_i\rangle$ have definite parity,

$$-\langle\phi_i|\hat{\bar{\mu}}_{ii}|\phi_i\rangle = 0, \tag{3.1.10}$$

so the matrix element is zero. We are left with the transition dipole moment matrix element between states 1 and 2, which may be measured and hence, is real. We define

$$\mu_{12} = \mu_{21} = \mu = \langle\phi_2|\hat{\bar{\mu}}_{21}|\phi_1\rangle, \tag{3.1.11}$$

and we return to the TDSE.

Our multiplication of Eq. (3.1.6) from the left by $\langle\phi_1|$ and then by $\langle\phi_2|$, leads to two coupled equations

$$i\hbar\frac{d}{dt}\begin{pmatrix} c_1(t)e^{-i\omega_1 t} \\ c_2(t)e^{-i\omega_2 t} \end{pmatrix} = \begin{pmatrix} \hbar\omega_1 & -\mu E\cos\omega t \\ -\mu E\cos\omega t & \hbar\omega_2 \end{pmatrix}\begin{pmatrix} c_1(t)e^{-i\omega_1 t} \\ c_2(t)e^{-i\omega_2 t} \end{pmatrix},$$

$$(3.1.12)$$

where we have used Eq. (3.1.5). The left-hand side of Eq. (3.1.12) becomes

$$\begin{pmatrix} i\hbar\left(\dfrac{dc_1(t)}{dt}\right)e^{-i\omega_1 t} + \hbar\omega_1 c_1(t)e^{-i\omega_1 t} \\ i\hbar\left(\dfrac{dc_2(t)}{dt}\right)e^{-i\omega_2 t} + \hbar\omega_2 c_2(t)e^{-i\omega_2 t} \end{pmatrix}. \qquad (3.1.13)$$

Now both the left-hand and the right-hand sides of Eq. (3.1.12) have $+\hbar\omega_i c_i(t)e^{-i\omega_i t}$, so we cancel these terms. We are left with the coupled equations

$$e^{-i\omega_1 t}\frac{dc_1(t)}{dt} = (i\mu E/\hbar)\cos\omega t c_2(t)e^{-i\omega_2 t}, \qquad (3.1.14)$$

$$e^{-i\omega_2 t}\frac{dc_2(t)}{dt} = (i\mu E/\hbar)\cos\omega t c_1(t)e^{-i\omega_1 t}, \qquad (3.1.15)$$

and we have shifted $i\hbar$ to the right-hand sides. Next, we define $\omega_0 = \omega_2 - \omega_1$ and rewrite the coupled equations as

$$\frac{dc_1(t)}{dt} = (i\mu E/\hbar)\cos\omega t c_2(t)e^{-i\omega_0 t}, \qquad (3.1.16)$$

$$\frac{dc_2(t)}{dt} = (i\mu E/\hbar)\cos\omega t c_1(t)e^{+i\omega_0 t}. \qquad (3.1.17)$$

The oscillations are completely exposed when we write the cosine in terms of exponentials

$$\frac{dc_1(t)}{dt} = (i\mu E/2\hbar)(e^{i(\omega-\omega_0)t} + e^{-i(\omega+\omega_0)t})c_2(t), \qquad (3.1.18)$$

$$\frac{dc_2(t)}{dt} = (i\mu E/2\hbar)(e^{i(\omega+\omega_0)t} + e^{-i(\omega-\omega_0)t})c_1(t). \qquad (3.1.19)$$

We first consider the weak-field limit, that is, when the electric field amplitude E is "small". This lets us determine some qualitative aspects of our two coupled equations. We start our system in level 1, so $c_1(t = 0) = 1$. and $c_2(t = 0) = 0$. The weak-field limit keeps $c_1(t) \approx 1$. and $c_2(t) \approx 0$. Thus, Eqs. (3.1.18) and (3.1.19) become

$$\frac{dc_1(t)}{dt} = 0, \tag{3.1.20}$$

$$\frac{dc_2(t)}{dt} = (i\mu E/2\hbar)(e^{i(\omega+\omega_0)t} + e^{-i(\omega-\omega_0)t}). \tag{3.1.21}$$

Equation (3.1.20) is consistent with $c_1(t) = 1$, while Eq. (3.1.21) has the solution

$$c_2(t) = (i\mu E/2\hbar)\left[\frac{e^{i(\omega+\omega_0)t} - 1}{i(\omega + \omega_0)} + \frac{e^{-i(\omega-\omega_0)t} - 1}{(-i)(\omega - \omega_0)}\right]. \tag{3.1.22}$$

This $c_2(t)$ is equal to 0 at $t = 0$. If we want the electromagnetic field to induce transitions between level 1 and level 2, then we need $\omega \approx \omega_0 = \omega_2 - \omega_1$. This lets us drop the first term in the square brackets of Eq. (3.1.22) and we have

$$c_2(t) \approx -(\mu E/2\hbar)(e^{-i(\omega-\omega_0)t} - 1)/(\omega - \omega_0). \tag{3.1.23}$$

With

$$\delta\omega = \omega - \omega_0, \tag{3.1.24}$$

a measurement finds the system in level 2 with a probability,

$$p_2(t) = c_2^*(t)c_2(t) \approx (\mu E/2\hbar)^2 2\frac{(1 - \cos\delta\omega t)}{(\delta\omega)^2}$$

$$= (\mu E/2\hbar)^2\frac{(\sin^2 \delta\omega t/2)}{(\delta\omega/2)^2}. \tag{3.1.25}$$

When $\delta\omega \approx 0$ or t is small,

$$p_2(t) \approx (\mu E/2\hbar)^2 t^2. \tag{3.1.26}$$

Hence, due to the weak electromagnetic field, the transition probability for going from level 1 to level 2 in this limit is

$$\frac{dp_2(t)}{dt} = 2(\mu E/2\hbar)^2 t. \tag{3.1.27}$$

Now the derivation of Einstein's A and B coefficients in Sec. 1.5 leads to the expectation that $p_2(t) \propto t$, so the transition probability is a constant. This dilemma is usually remedied by invoking a line width for the transition and/or a spectral energy density for the electromagnetic field. An integral over energy or frequency then gives $p_2(t) \propto t$ (Fox, 2006, Sec. 9.4).

This ends our weak-field digression. Further progress requires numerical approaches or the RWA. We take the latter path in Sec. 3.2.

3.2. The Rotating Wave Approximation Enters

We return to Eqs. (3.1.18) and (3.1.19), remove the weak-field assumption and rewrite the equations in terms of

$$\delta\omega = \omega - \omega_0 = \omega - (\omega_2 - \omega_1), \tag{3.2.1}$$

hence,

$$\frac{dc_1(t)}{dt} = (i\mu E/2\hbar)(e^{i\delta\omega t} + e^{-i(\omega+\omega_0)t})c_2(t), \tag{3.2.2}$$

$$\frac{dc_2(t)}{dt} = (i\mu E/2\hbar)(e^{-i\delta\omega t} + e^{+i(\omega+\omega_0)t})c_1(t). \tag{3.2.3}$$

If $|\delta\omega| \ll \omega_0$, then the exponentials with $\omega + \omega_0$ vary rapidly compared to the exponentials with $\delta\omega$. So, it is reasonable to assume the rapidly varying terms contribute little when Eqs. (3.2.2) and (3.2.3) are integrated. Thus, we drop such terms and, in doing so, we make the Rotating Wave Approximation (RWA). We define

$$\tilde{\Omega} = \mu E/\hbar, \tag{3.2.4}$$

and write the coupled equations in the RWA as

$$\frac{dc_1(t)}{dt} = (i\tilde{\Omega}/2)e^{i\delta\omega t}c_2(t), \tag{3.2.5}$$

$$\frac{dc_2(t)}{dt} = (i\tilde{\Omega}/2)e^{-i\delta\omega t}c_1(t). \tag{3.2.6}$$

We remark that some authors refer to $\tilde{\Omega}$ as the Rabi frequency or the Rabi angular frequency. We follow the convention where the Rabi angular frequency is Ω and

$$\Omega = \sqrt{(\delta\omega)^2 + \tilde{\Omega}^2}. \tag{3.2.7}$$

Our next task is to solve the RWA coupled equations and see how Ω arises.

An examination of the coupled equations hints that we should take the time derivative of the first equation, Eq. (3.2.5), and then substitute the second equation, Eq. (3.2.6), to eliminate the first derivative of $c_2(t)$. This approach leads to a second-order equation for $c_1(t)$ when $c_2(t)$ is replaced by Eq. (3.2.5). We start with

$$\frac{d^2 c_1(t)}{dt^2} = (i\tilde{\Omega} i\delta\omega/2)e^{i\delta\omega t}c_2(t) + (i\tilde{\Omega}/2)e^{i\delta\omega t}\frac{dc_2(t)}{dt}. \tag{3.2.8}$$

Now we substitute Eq. (3.2.6) for dc_2/dt and use Eq. (3.2.5) to replace $c_2(t)$,

$$\frac{d^2 c_1(t)}{dt^2} = i\delta\omega \frac{dc_1(t)}{dt} + (i\tilde{\Omega}/2)e^{i\delta\omega t}(i\tilde{\Omega}/2)e^{-i\delta\omega t}c_1(t). \tag{3.2.9}$$

This last equation is rearranged to show the second-order ordinary differential equation for $c_1(t)$

$$\frac{d^2 c_1(t)}{dt^2} - i\delta\omega \frac{dc_1(t)}{dt} + (\tilde{\Omega}/2)^2 c_1(t) = 0. \tag{3.2.10}$$

We try an exponential solution for $c_1(t)$,

$$c_1(t) = c_1 e^{i\alpha t}. \tag{3.2.11}$$

We substitute our guess into Eq. (3.2.10) and find a quadratic equation for α,

$$\alpha^2 - (\delta\omega)\alpha - (\tilde{\Omega}/2)^2 = 0. \tag{3.2.12}$$

The solutions are

$$\alpha_\pm = (\delta\omega/2) \pm (1/2)\sqrt{(\delta\omega)^2 + (\tilde{\Omega})^2} = \frac{\delta\omega}{2} \pm \frac{\Omega}{2}, \tag{3.2.13}$$

and we see the emergence of the Rabi angular frequency Ω. Thus, our solution for $c_1(t)$ is

$$c_1(t) = c_{1+}e^{i\delta\omega t/2}e^{i\Omega t/2} + c_{1-}e^{i\delta\omega t/2}e^{-i\Omega t/2}, \qquad (3.2.14)$$

with the constants c_{1+} and c_{1-} left to determine.

We follow a similar procedure to get a second-order ordinary differential equation for $c_2(t)$. We start with

$$\frac{d^2 c_2(t)}{dt^2} = -(i\tilde{\Omega}i\delta\omega/2)e^{-i\delta\omega t}c_1(t) + (i\tilde{\Omega}/2)e^{-i\delta\omega t}\frac{dc_1(t)}{dt}, \qquad (3.2.15)$$

then we substitute for the $c_1(t)$ terms with Eqs. (3.2.5) and (3.2.6) and find

$$\frac{d^2 c_2(t)}{dt^2} = -i\delta\omega\frac{dc_2(t)}{dt} + (i\tilde{\Omega}/2)e^{-i\delta\omega t}(i\tilde{\Omega}/2)e^{+i\delta\omega t}c_2(t). \qquad (3.2.16)$$

And this equation simplifies to

$$\frac{d^2 c_2(t)}{dt^2} + i\delta\omega\frac{dc_2(t)}{dt} + (\tilde{\Omega}/2)^2 c_2(t) = 0. \qquad (3.2.17)$$

Now we take

$$c_2(t) = c_2 e^{-i\beta t}, \qquad (3.2.18)$$

and please note the minus sign. This choice provides the needed cancellations when we put the solutions for $c_1(t)$ and $c_2(t)$ into Eqs. (3.2.5) and (3.2.6), and ensures the solutions are consistent. When Eq. (3.2.18) is placed into Eq. (3.2.17), we find

$$\beta^2 - (\delta\omega)\beta - (\tilde{\Omega}/2)^2 = 0, \qquad (3.2.19)$$

and

$$\beta_\pm = (\delta\omega/2) \pm (1/2)\sqrt{(\delta\omega)^2 + (\tilde{\Omega})^2} = \frac{\delta\omega}{2} \pm \frac{\Omega}{2}. \qquad (3.2.20)$$

Finally, we have

$$c_2(t) = c_{2+}e^{-i\delta\omega t/2}e^{-i\Omega t/2} + c_{2-}e^{-i\delta\omega t/2}e^{+i\Omega t/2}. \qquad (3.2.21)$$

Equations (3.2.14) and (3.2.21) are linked by the coupled Eqs. (3.2.5) and (3.2.6), and it suffices to work with either equation of the latter two. We choose Eq. (3.2.5) and find

$$
c_{1+}\left(\frac{i\delta\omega}{2}+\frac{i\Omega}{2}\right)e^{i\delta\omega t/2}e^{i\Omega t/2} + c_{1-}\left(\frac{i\delta\omega}{2}-\frac{i\Omega}{2}\right)e^{i\delta\omega t/2}e^{-i\Omega t/2}
$$
$$
= i\frac{\tilde{\Omega}}{2}e^{i\delta\omega t}\left(c_{2+}e^{-i\delta\omega t/2}e^{-i\Omega t/2} + c_{2-}e^{-i\delta\omega t/2}e^{+i\Omega t/2}\right).
$$
$$(3.2.22)$$

We cancel $i/2$ from each term and also $e^{i\delta\omega t/2}$ due to our sign choice in Eq. (3.2.18). If Eq. (3.2.22) is to hold for all times t, we need to match terms. The $e^{i\Omega t/2}$ terms lead to

$$
(\delta\omega + \Omega)c_{1+} = \tilde{\Omega}c_{2-}, \tag{3.2.23}
$$

and the $e^{-i\Omega t/2}$ terms provide

$$
(\delta\omega - \Omega)c_{1-} = \tilde{\Omega}c_{2+}. \tag{3.2.24}
$$

We now need the initial conditions. We assume the system starts in level 1, the lower energy level,

$$
c_1(t=0) = 1, \tag{3.2.25}
$$
$$
c_2(t=0) = 0. \tag{3.2.26}
$$

When the last equation is applied to Eq. (3.2.21) at $t = 0$, we find

$$
c_{2+} = -c_{2-}. \tag{3.2.27}
$$

Then Eq. (3.2.14) at $t = 0$ with Eq. (3.2.25) yields

$$
1 = c_{1+} + c_{1-} = \left(\frac{-\tilde{\Omega}}{\delta\omega + \Omega} + \frac{\tilde{\Omega}}{\delta\omega - \Omega}\right)c_{2+}, \tag{3.2.28}
$$

when Eqs. (3.2.23), (3.2.24) and (3.2.27) are used. Equation (3.2.28) simplifies to

$$
c_{2+} = -\frac{\tilde{\Omega}}{2\Omega}, \tag{3.2.29}
$$

through the use of the definition of Ω from Eq. (3.2.7). We get the $c_{1\pm}$ with Eqs. (3.2.23) and (3.2.24)

$$c_{1+} = \frac{\tilde{\Omega}^2}{2\Omega(\delta\omega + \Omega)}, \tag{3.2.30}$$

$$c_{1-} = \frac{-\tilde{\Omega}^2}{2\Omega(\delta\omega - \Omega)}. \tag{3.2.31}$$

Thus, for the initial conditions given in Eqs. (3.2.25) and (3.2.26), we finally arrive at

$$c_1(t) = \frac{\tilde{\Omega}^2}{2\Omega}\left(\frac{e^{i\delta\omega t/2}e^{i\Omega t/2}}{\delta\omega + \Omega} - \frac{e^{i\delta\omega t/2}e^{-i\Omega t/2}}{\delta\omega - \Omega}\right), \tag{3.2.32}$$

$$c_2(t) = \frac{\tilde{\Omega}}{2\Omega}\left(-e^{-i\delta\omega t/2}e^{-i\Omega t/2} + e^{-i\delta\omega t/2}e^{i\Omega t/2}\right). \tag{3.2.33}$$

These expressions may be simplified by grouping exponentials into cosine and sine,

$$c_1(t) = e^{i\delta\omega t/2}\left[\cos(\Omega t/2) - \frac{i\delta\omega}{\Omega}\sin(\Omega t/2)\right], \tag{3.2.34}$$

$$c_2(t) = e^{-i\delta\omega t/2}\left[\frac{i\tilde{\Omega}}{\Omega}\sin(\Omega t/2)\right]. \tag{3.2.35}$$

And these expressions lead to the occupation probabilities of each level as a function of time

$$p_1(t) = |c_1(t)|^2 = [\cos(\Omega t/2)]^2 + \left(\frac{\delta\omega}{\Omega}\right)^2 [\sin(\Omega t/2)]^2, \tag{3.2.36}$$

$$p_2(t) = |c_2(t)|^2 = \left(\frac{\tilde{\Omega}}{\Omega}\right)^2 [\sin(\Omega t/2)]^2. \tag{3.2.37}$$

And we have Rabi oscillations! These are periodic oscillations in the occupation probabilities. As a check, we use Eq. (3.2.7) yet again and check the sum of the occupation probabilities,

$$p_1(t) + p_2(t) = |c_1(t)|^2 + |c_2(t)|^2$$

$$= \cos^2(\Omega t/2) + \frac{(\delta\omega)^2 + (\tilde{\Omega})^2}{\Omega^2} \sin^2(\Omega t/2) = 1.$$

$$(3.2.38)$$

So, the total occupation probability stays at 1, which is reassuring.

It is worth pointing out that the $c_i(t)$ depend nonlinearly on the strength of the electromagnetic field strength through $\tilde{\Omega} = \mu E/\hbar$. This means a superposition of coupling potentials does not lead to a superposition of solutions linear in E.

If the electromagnetic field's angular frequency ω equals the separation of the energy levels ω_0, then we have resonance and $\delta\omega = 0$. In the case of resonance, the $c_i(t)$ of Eqs. (3.2.34) and (3.2.35) become

$$c_1(t) = \cos(\tilde{\Omega}t/2), \qquad (3.2.39)$$

$$c_2(t) = i\sin(\tilde{\Omega}t/2), \qquad (3.2.40)$$

and the resulting $p_i(t)$ are

$$p_1(t) = |c_1(t)|^2 = [\cos(\tilde{\Omega}t/2)]^2, \qquad (3.2.41)$$

$$p_2(t) = |c_2(t)|^2 = [\sin(\tilde{\Omega}t/2)]^2. \qquad (3.2.42)$$

The following section has plots of the above occupation probabilities and the $c_i(t)$ for the case when the system starts in the upper energy level.

3.3. Exploring the Solutions for a Two-Level System

We continue with a discussion of the solutions developed in Sec. 3.2 for a coupling potential operator $\hat{V}_c$ that is proportional to $\cos\omega t$ as in Eq. (3.1.1). We started the system in level 1 and found the occupation probability, $p_2(t)$, for the system to be in level 2 at time t to be given by Eq. (3.2.37). When the system is off-resonance, that is, when

$$\delta\omega = \omega - \omega_0 = \omega - (\omega_2 - \omega_1) \neq 0, \qquad (3.3.1)$$

then $p_2(t)$ has a maximum less than 1. This follows from

$$\tilde{\Omega} < \Omega = \sqrt{(\delta\omega)^2 + \tilde{\Omega}^2}. \qquad (3.3.2)$$

Here $\tilde{\Omega}$ is defined in Eq. (3.2.4) and is proportional to the magnitude of the electromagnetic field. Thus, when the system is off-resonance, the occupation probability $p_1(t)$ of level 1 never drops to 0. In addition, since the $p_i(t)$ depend on Ω, the $p_i(t)$ oscillate at a higher frequency than the $p_i(t)$ for the resonance case with $\delta\omega = 0$, which depend only on $\tilde{\Omega}$. When resonance applies, then $p_2(t)$ reaches 1 and $p_1(t)$ falls to 0. This happens when $\tilde{\Omega}t/2$ is an odd integer multiple of $\pi/2$. Both the resonance and the off-resonance cases lead to the short-time behavior of

$$p_2(t) \rightarrow t^2. \qquad (3.3.3)$$

Now, we use plots to illustrate the above points and to show Rabi oscillations. We reiterate that we are using the Rotating Wave Approximation (RWA). The $p_i(t)$ give the probability of a measurement finding the two-level system in level i at time t. Sometimes, if we envision an ensemble of identical two-level systems, we may speak in terms of the populations of the levels or the number of systems in a given level. We use the $p_i(t)$ in this sense also. We continue to assume we start with $p_1(t = 0) = 1$. We first plot $p_1(t)$ and $p_2(t)$ for the resonance case with $\delta\omega = 0$ in Fig. 3.1. Here the time scale is the inverse of the angular frequency scale and we use Eqs. (3.2.41) and (3.2.42). The complete emptying of level 1 and the accompanying transfer of the population to level 2 is shown, as well as the continuing oscillations of the populations.

Suppose with $\delta\omega = 0$, we suddenly turn the electromagnetic field off. This may be viewed as a pulse. When this occurs at time t such that $\tilde{\Omega}t = \pi$, then $p_1 = 0$ and $p_2 = 1$, and the entire population is now in the upper level. Such a pulse is known as a π-pulse. If the pulse duration is half this time, then $p_1 = p_2 = 0.5$, and this is

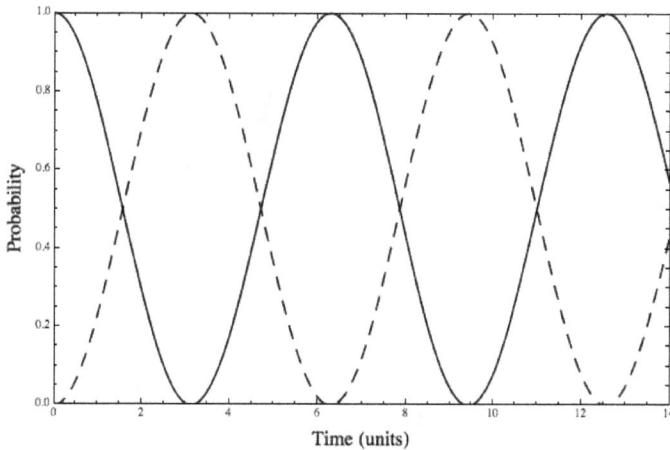

Figure 3.1. Occupation probabilities for a two-level system at resonance in the RWA with $\tilde{\Omega} = 1$. The system starts in level 1, the lower energy level. $p_1(t)$ is solid and $p_2(t)$ is dashed.

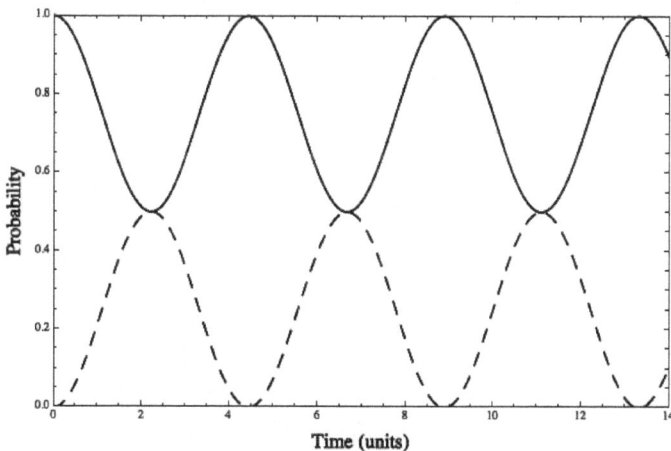

Figure 3.2. Occupation probabilities for a two-level system with $\delta\omega = 1$. The system starts in level 1, the lower energy level. $p_1(t)$ is solid, $p_2(t)$ is dashed and $\tilde{\Omega} = 1$. The RWA is used.

known as a $\pi/2$-pulse. We return to these pulses in Chap. 4 when we consider magnetic resonance.

When $\delta\omega \neq 0$, we continue to use the same time scale and we specify $\delta\omega$ and $\tilde{\Omega}$. Figures 3.2 to 3.4 are for $\delta\omega = 1.0, 1.5$ and 2.0,

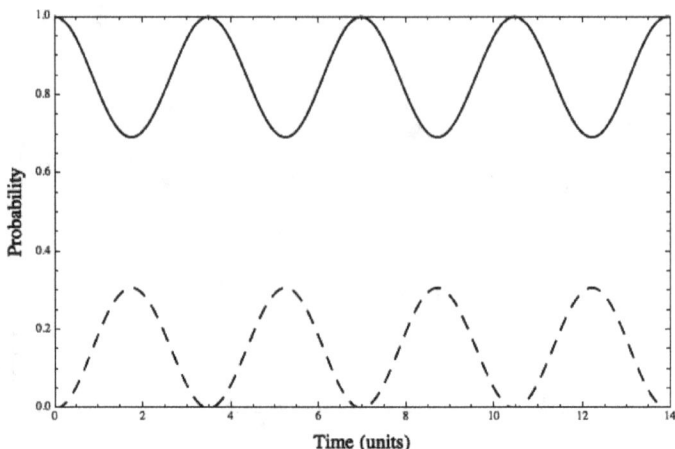

Figure 3.3. The RWA occupation probabilities for a two-level system with $\delta\omega = 1.5$. The system starts in level 1, the lower energy level. $p_1(t)$ is solid, $p_2(t)$ is dashed and $\tilde{\Omega} = 1$.

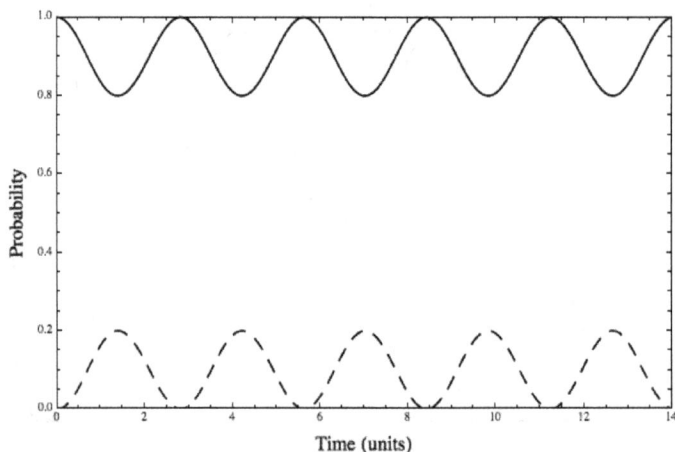

Figure 3.4. Occupation probabilities for a two-level system with $\delta\omega = 2$. The system starts in level 1, the lower energy level. $p_1(t)$ is solid, $p_2(t)$ is dashed and $\tilde{\Omega} = 1$. The RWA is used.

respectively. We set $\tilde{\Omega} = 1$ and we use Eqs. (3.2.36) and (3.2.37), which are the solutions in the RWA. We note that as the frequency ω of the electromagnetic field gets further away from $\omega_0 = \omega_2 - \omega_1$, the occupation probability of level 2 reaches a lower maximum

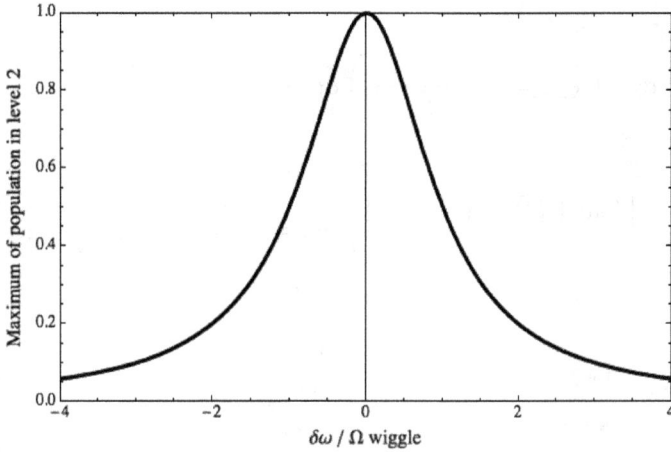

Figure 3.5. Maximum of $p_2(t)$ versus $\delta\omega/\tilde{\Omega}$. Here $\delta\omega$ is defined in Eq. (3.3.1) and $\tilde{\Omega} = \Omega$ wiggle. The system starts in level 1, the lower energy level and the RWA is used.

value. This means the electromagnetic field is less likely to induce a transition from level 1 to level 2. On the other hand, for a fixed $\delta\omega$, the maximum occupation probability of level 2 increases with an increase in $\tilde{\Omega}$ which is proportional to the electromagnetic field magnitude. This situation is in accord with Eqs. (3.2.36) and (3.2.37). Finally, we plot the maximum of $p_2(t)$ versus $\delta\omega$ in Fig. 3.5, which is a curve of $(\tilde{\Omega}/\Omega)^2$. The result is a Lorentzian with the full width at half-maximum equal to twice $\tilde{\Omega}$.

We next derive the occupation probabilities when we start the system in the upper level, level 2. We parallel the derivation in the second half of Sec. 3.2 from Eq. (3.2.23) onward.

Now the initial conditions are

$$c_1(t = 0) = 0, \tag{3.3.4}$$

$$c_2(t = 0) = 1. \tag{3.3.5}$$

So,

$$c_{1-} = -c_{1+}, \tag{3.3.6}$$

and

$$1 = c_{2+} + c_{2-} = \left(\frac{(\delta\omega - \Omega)}{\tilde{\Omega}} \right) c_{1-} + \left(\frac{(\delta\omega + \Omega)}{\tilde{\Omega}} \right) c_{1+}. \qquad (3.3.7)$$

Hence,

$$\left[\frac{(\delta\omega + \Omega)}{\tilde{\Omega}} - \frac{(\delta\omega - \Omega)}{\tilde{\Omega}} \right] c_{1+} = \frac{2\Omega}{\tilde{\Omega}} c_{1+} = 1, \qquad (3.3.8)$$

and

$$c_{1+} = \frac{\tilde{\Omega}}{2\Omega}, \qquad (3.3.9)$$

$$c_{1-} = -\frac{\tilde{\Omega}}{2\Omega}. \qquad (3.3.10)$$

With Eqs. (3.2.23) and (3.2.24), these in turn lead to

$$c_{2+} = \left(\frac{\delta\omega - \Omega}{\tilde{\Omega}} \right) c_{1-} = -\frac{(\delta\omega - \Omega)}{2\Omega}, \qquad (3.3.11)$$

$$c_{2-} = \left(\frac{\delta\omega + \Omega}{\tilde{\Omega}} \right) c_{1+} = \frac{(\delta\omega + \Omega)}{2\Omega}. \qquad (3.3.12)$$

Equation (3.2.14) leads us to $c_1(t)$ for the initial conditions of Eqs. (3.3.4) and (3.3.5)

$$c_1(t) = \frac{\tilde{\Omega}}{2\Omega} (e^{i\delta\omega t/2} e^{i\Omega t/2} - e^{i\delta\omega t/2} e^{-i\Omega t/2}). \qquad (3.3.13)$$

Similarly, Eq. (3.2.21) brings us to

$$c_2(t) = \frac{1}{2\Omega} (-(\delta\omega - \Omega) e^{-i\delta\omega t/2} e^{-i\Omega t/2}$$
$$+ (\delta\omega + \Omega) e^{-i\delta\omega t/2} e^{+i\Omega t/2}). \qquad (3.3.14)$$

These equations for $c_1(t)$ and $c_2(t)$ are simplified to

$$c_1(t) = \frac{i\tilde{\Omega}}{\Omega} e^{i\delta\omega t/2} \sin(\Omega t/2), \qquad (3.3.15)$$

$$c_2(t) = e^{-i\delta\omega t/2} \left(\frac{i\delta\omega}{\Omega} \sin(\Omega t/2) + \cos(\Omega t/2) \right). \qquad (3.3.16)$$

Finally, the occupation probabilities are

$$p_1(t) = |c_1(t)|^2 = \left(\frac{\tilde{\Omega}}{\Omega}\right)^2 [\sin(\Omega t/2)]^2, \qquad (3.3.17)$$

and

$$p_2(t) = |c_2(t)|^2 = [\cos(\Omega t/2)]^2 + \left(\frac{\delta\omega}{\Omega}\right)^2 [\sin(\Omega t/2)]^2. \qquad (3.3.18)$$

When we compare these occupation probabilities with those of Eqs. (3.2.36) and (3.2.37), we see that they are interchanged when we start the system in level 2 instead of in level 1.

3.3.1. *RWA Solutions versus Numerical Solutions*

Before we continue, it is useful to compare the above Rotating Wave Approximation (RWA) solutions with the numerical solutions of Eqs. (3.1.16) and (3.1.17). The latter do not make the RWA and this allows us to determine the influence of the terms with $\omega + \omega_0$. Here ω is the angular frequency of the applied electromagnetic field and ω_0 is the angular frequency difference between the two energy levels. We expect to see high frequency terms in the plots for $p_i(t)$ due to the retention of the sum terms. We assume the population starts in the lower energy level, so we use the initial conditions given by Eqs. (3.2.25) and (3.2.26). The numerical solutions are obtained with the routine NDSolve of Mathematica version 9 (2012). The RWA results are provided by Eqs. (3.2.36) and (3.2.37) and $\tilde{\Omega}$ is defined by Eq. (3.2.4).

The initial set of plots use $\tilde{\Omega} = 0.5$ and explore resonance. Figures 3.6, 3.7 and 3.8 have the angular frequencies $\omega = \omega_0$ set to 1, 2 and 4, respectively. The numerical solutions of Eqs. (3.1.16) and (3.1.17) are referred to here as the full solutions and are plotted along with the RWA solutions. As the angular frequency increases, the oscillating overlay of the full solution increases its frequency in step. In addition, we see the full solutions more closely resemble the RWA solutions, because the higher frequency components decrease in magnitude with an increase in ω. This may be shown by an

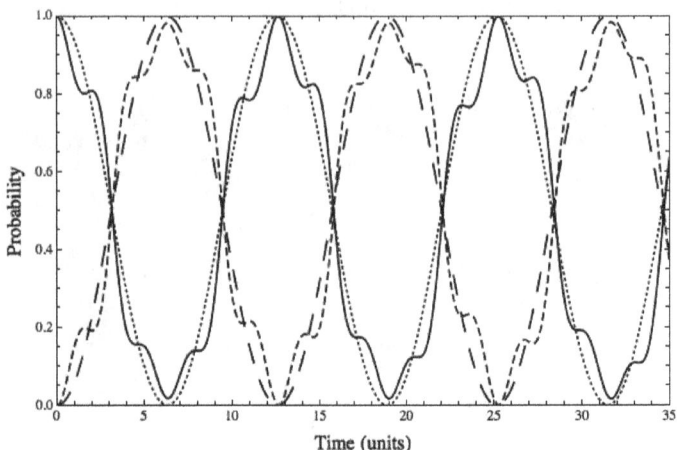

Figure 3.6. Occupation probabilities for a two-level system with $\omega = \omega_0 = 1$ and $\tilde{\Omega} = 0.5$. The system starts in level 1, the lower energy level. $p_1(t)$ is solid and $p_2(t)$ is short-dashed for the full solution. $p_1(t)$ is dotted and $p_2(t)$ is long-dashed for the RWA solution.

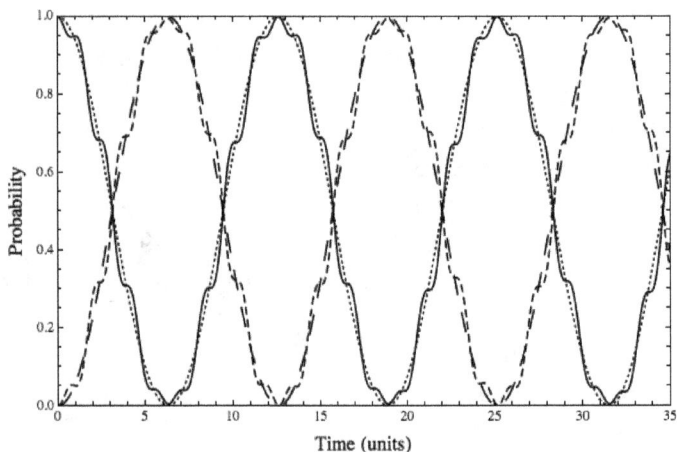

Figure 3.7. Occupation probabilities for a two-level system with $\omega = \omega_0 = 2$ and $\tilde{\Omega} = 0.5$. The system starts in level 1, the lower energy level. $p_1(t)$ is solid and $p_2(t)$ is short-dashed for the full solution. $p_1(t)$ is dotted and $p_2(t)$ is long-dashed for the RWA solution.

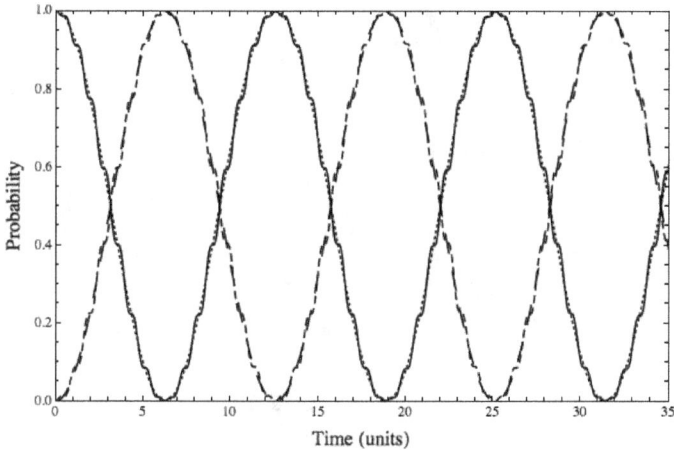

Figure 3.8. Occupation probabilities for a two-level system with $\omega = \omega_0 = 4$ and $\tilde{\Omega} = 0.5$. The system starts in level 1, the lower energy level. $p_1(t)$ is solid and $p_2(t)$ is short-dashed for the full solution. $p_1(t)$ is dotted and $p_2(t)$ is long-dashed for the RWA solution.

approximate argument that follows Bonacci (2004) and is presented in App. 5. The result is the higher frequency terms in $c_i(t)$ go as $\tilde{\Omega}/(\omega + \omega_0)$ and, hence, decrease when the denominator increases. The peaks in both sets of solutions agree reasonably well.

We go off-resonance in Fig. 3.9 with $\omega_0 = 4.0$ and $\omega = 4.2$ and compare this figure with Fig. 3.8. Firstly, we note that in both figures, the full solutions show higher frequency components on top of the RWA solutions of Eqs. (3.2.36) and (3.2.37). But the high frequency components now overlay the RWA curves less than in Fig. 3.8. Secondly, with an increase in time, the full solutions are slowly edging to the right of their RWA counterparts. The full solutions require a longer time to reach an extrema. We interpret this shift by assuming the $p_i(t)$ depend on a function of $\eta = (\tilde{\Omega}^2 + (\delta\omega)^2)^{1/2}t$ and we draw on Eq. (3.2.1). Effectively, we suspect for $\omega > \omega_0$, as in Fig. 3.9, that it is as if ω_0 is increasing due to the electric field. Sie *et al.* (2017) have seen this in experiments and attributed the shift to the optical Stark shift and the Bloch–Siegert shift. The increase in ω_0 leads to a smaller $\delta\omega$ and a larger time t to give η the same value. We remark that for $\omega = 3.8$ the extrema of the full solution shift to the left of the

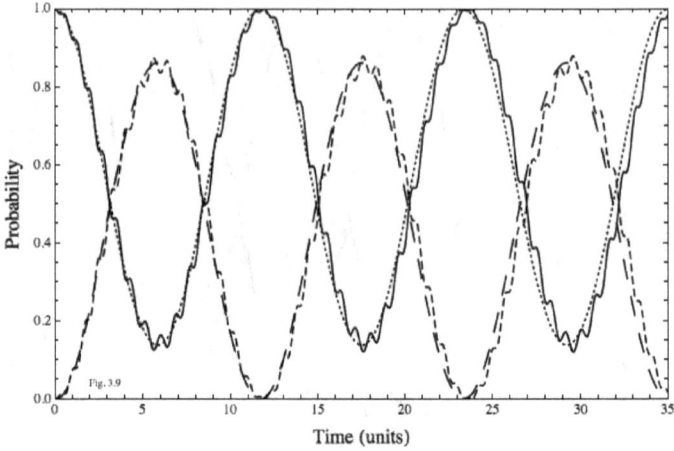

Figure 3.9. Occupation probabilities for a two-level system with $\omega = 4.2$, $\omega_0 = 4$ and $\tilde{\Omega} = 0.5$. The system starts in level 1, the lower energy level. $p_1(t)$ is solid and $p_2(t)$ is short-dashed for the full solution. $p_1(t)$ is dotted and $p_2(t)$ is long-dashed for the RWA solution. The extrema of the full solution are shifting to longer times.

extrema of the RWA solution. Here the larger ω_0 leads to an increase in $|\delta\omega|$ and a smaller time t is needed to maintain η.

The question of when the RWA suffices is often touched upon in writings on Quantum Optics. However, definitive statements are hard to locate. Perrin (2014) states the RWA holds if $\delta\omega$ and $\tilde{\Omega}$ are both much less than ω and ω_0. Ketterle (2014) says the RWA is valid when $|\omega - \omega_0| \ll \omega_0$ and $\tilde{\Omega} \ll \omega_0$. We see that Figs. 3.6 to 3.8 are successively more appropriate for the use of the RWA. Figure 3.9 shows that a departure from resonance is likely to require the full solution for an accurate treatment.

Bonacci (2004) also discusses changes such as those we see in Figs. 3.6 to 3.9, but from an alternative viewpoint. Equations (3.1.16) and (3.1.17) are turned into two coupled second-order linear differential equations to expose which terms dominate, and when. Details are in App. 5. It is helpful to go to dimensionless variables such as

$$\tau = \tilde{\Omega}t, \qquad\qquad (3.3.19)$$

$$\delta = (\omega - \omega_0)/\tilde{\Omega}, \qquad\qquad (3.3.20)$$

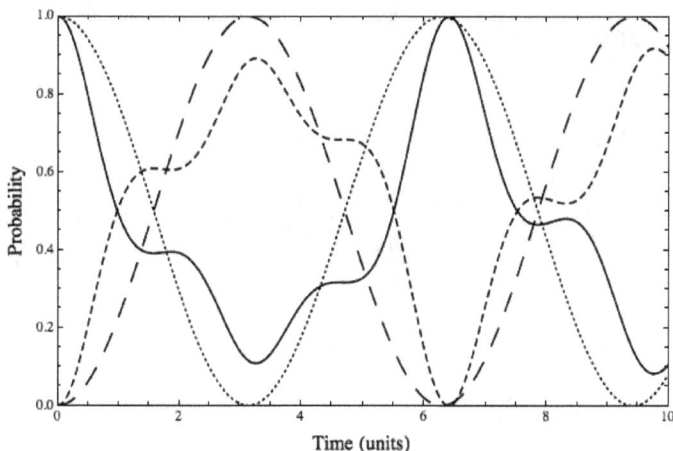

Figure 3.10.   Occupation probabilities for a two-level system with $\omega = 1$, $\omega_0 = 1$ and $\tilde{\Omega} = 1.0$. The system starts in level 1, the lower energy level. $p_1(t)$ is solid and $p_2(t)$ is short-dashed for the full solution. $p_1(t)$ is dotted and $p_2(t)$ is long-dashed for the RWA solution.

and

$$\gamma = (\omega + \omega_0)/\tilde{\Omega}. \tag{3.3.21}$$

Each second-order equation has a term with δ and a term with γ multiplying an oscillatory function of time as well as two other time-dependent terms. These terms are developed in App. 5. Bonacci finds this approach useful for making qualitative arguments such as that the solution's behavior is different for $\gamma \gg 1$ and for $\gamma \ll 1$. Figures 3.6 to 3.9 are for $\gamma > 1$ and reach $\gamma = 16.4$. The behavior we see includes the superposed oscillations and the slight shift in the locations of the $p_i(t)$ extrema that Bonacci describes. Further examples for $\gamma > 1$ are included in App. 5 and Sec. 3.3.2 shows data for $\gamma \gg 1$.

The other limit, $\gamma \ll 1$, produces more unusual features as shown by the two lower plots in Bonacci (2004, Fig. 2). We now venture into this region with $\omega = \omega_0 = 1$ and vary the field strength or the dipole moment, since their product goes into $\tilde{\Omega}$ according to Eq. (3.2.4). Figures 3.10 to 3.12 have $\tilde{\Omega}$ equal to 1.0, 1.5 and 2.0, respectively. Figure 3.6 provides a fourth plot to start this series with $\tilde{\Omega} = 0.5$.

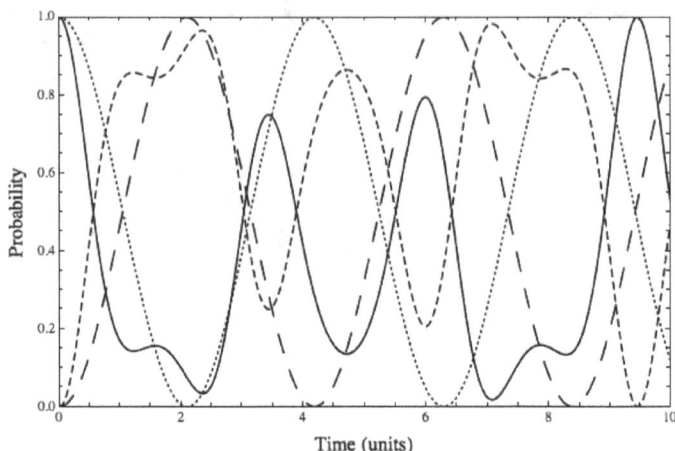

Figure 3.11. Occupation probabilities for a two-level system with $\omega = 1$, $\omega_0 = 1$ and $\tilde{\Omega} = 1.5$. The system starts in level 1, the lower energy level. $p_1(t)$ is solid and $p_2(t)$ is short-dashed for the full solution. $p_1(t)$ is dotted and $p_2(t)$ is long-dashed for the RWA solution.

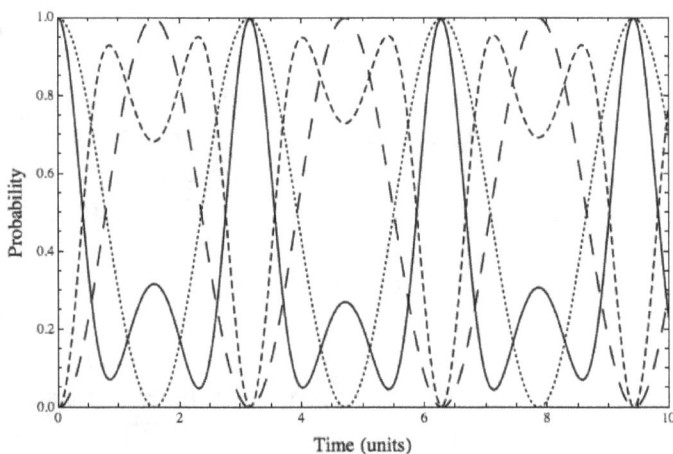

Figure 3.12. Occupation probabilities for a two-level system with $\omega = 1$, $\omega_0 = 1$ and $\tilde{\Omega} = 2.0$. The system starts in level 1, the lower energy level. $p_1(t)$ is solid and $p_2(t)$ is short-dashed for the full solution. $p_1(t)$ is dotted and $p_2(t)$ is long-dashed for the RWA solution.

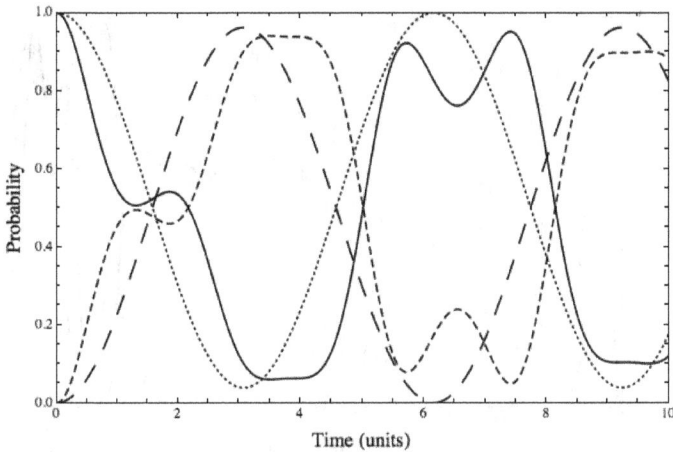

Figure 3.13. Occupation probabilities for a two-level system with $\omega = 1.2$, $\omega_0 = 1$ and $\tilde{\Omega} = 1.0$. The system starts in level 1, the lower energy level. $p_1(t)$ is solid and $p_2(t)$ is short-dashed for the full solution. $p_1(t)$ is dotted and $p_2(t)$ is long-dashed for the RWA solution.

The increase in $\tilde{\Omega}$ brings forth full solutions that differ markedly from the RWA solutions. We see we are violating the condition $\tilde{\Omega} \ll \omega_0$ with these cases, so we should not be surprised to find that the RWA is inadequate. Additional peaks and minima now occur in the $p_i(t)$ for the full solutions and complete periodicity is not evident on the time scale of the plots.

Figure 3.13 introduces a non-resonant case with $\omega_0 = 1$, $\omega = 1.2$ and $\tilde{\Omega} = 1$. A comparison with Fig. 3.10 shows dramatic changes in the $p_i(t)$ for the full solution. This case is extended to longer times in Fig. 3.14. It appears that each significant change in the slope of a $p_i(t)$ evolves over the time plotted. Clearly, there are plenty of challenges in understanding the $\gamma < 1$ regime!

Further examples of the full solutions for the $p_i(t)$ are presented in Sec. 4.4 along with a comparison to published data for a wide range of field strengths (Fuchs *et al.*, 2009). We next enter Sec. 3.3.2 and examine a recent measurement of Rabi oscillations for $\gamma \gg 1$.

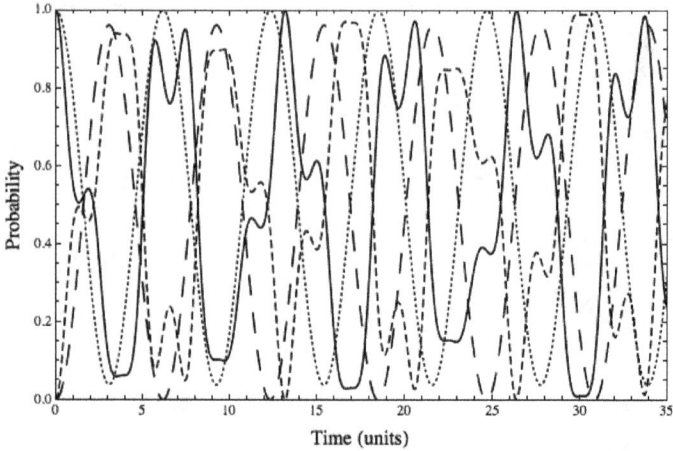

Figure 3.14. Occupation probabilities for a two-level system with $\omega = 1.2$, $\omega_0 = 1$ and $\tilde{\Omega} = 1.0$ for longer times than in Fig. 3.13. The system starts in level 1, the lower energy level. $p_1(t)$ is solid and $p_2(t)$ is short-dashed for the full solution. $p_1(t)$ is dotted and $p_2(t)$ is long-dashed for the RWA solution.

3.3.2. *An Example of Rabi Oscillations*

This section contains a recent example of Rabi oscillations observed between the hyperfine levels of the $6S^1$ state of a trapped ytterbium (Yb) ion. While Chap. 4 presents several examples of time dependence in two-level systems, it is worthwhile to first illustrate the ideas presented so far with the experimental observation of Rabi oscillations. Trapped Yb ions are used for quantum information processing and quantum simulation of physical systems (Roos, 2014). Olmschenk *et al.* (2007) use a Paul trap to arrange ^{171}Yb$^+$ ions in a linear chain, but they also present data for an individual ^{171}Yb$^+$. We find the Hamiltonian for one ion and then the associated time-dependent equations for the wave functions. Finally, we compare the data to our calculations within the Rotating Wave Approximation (RWA).

The relevant state of the ^{171}Yb$^+$ ion is $^2S_{1/2}$ and we have the same hyperfine levels as found in the hydrogen atom discussed in Chap. 2, Ramble 3. We use the results developed there to deduce the Time-Dependent Schrödinger Equation for the ^{171}Yb$^+$ hyperfine levels. We need the total angular momentum F, which depends here on the spin of the nucleus $I = 1/2$ (Enge, 1966, App. 6) and the spin of the electron $S = 1/2$. The transitions occur between the ground state $|F = 0, m_F = 0\rangle$ and the excited state $|F = 1, m_F = 0\rangle$. The measured frequency difference is 12.6428 GHz. We take the Hamiltonian

$$\hat{H} = A\hat{\vec{I}} \cdot \hat{\vec{S}} - \hat{\vec{M}}_S \cdot \vec{B}, \tag{3.3.22}$$

for the electron with

$$-\hat{\vec{M}}_S \cdot \vec{B}(t) = \frac{g\mu_B B}{\hbar}\hat{S}_z \cos \omega t = \Omega_0 \hat{S}_z \cos \omega t. \tag{3.3.23}$$

The numerical value of the A in Eq. (3.2.22) is different from its value in Ramble 3.

We next determine the action of $\hat{S}_z$ on the hyperfine states. We write the $|F, m_F\rangle$ in terms of the $|S_Z, I_Z\rangle$ states and start with

$$\hat{S}_z|1, \pm 1\rangle = \hat{S}_z \left|\pm\frac{1}{2}, \pm\frac{1}{2}\right\rangle = \pm\frac{\hbar}{2}\left|\pm\frac{1}{2}, \pm\frac{1}{2}\right\rangle = \pm\frac{\hbar}{2}|1, \pm 1\rangle. \tag{3.3.24}$$

We see the $\hat{S}_z$ term does not couple either of these states to another state. However, the two $m_F = 0$ states are coupled. First,

$$\hat{S}_z|1, 0\rangle = \hat{S}_z \frac{1}{\sqrt{2}}\left(\left|\frac{1}{2}, -\frac{1}{2}\right\rangle + \left|-\frac{1}{2}, \frac{1}{2}\right\rangle\right)$$

$$= \frac{\hbar}{2\sqrt{2}}\left(\left|\frac{1}{2}, -\frac{1}{2}\right\rangle - \left|-\frac{1}{2}, \frac{1}{2}\right\rangle\right) = \frac{\hbar}{2}|0, 0\rangle, \tag{3.3.25}$$

and similarly,

$$\hat{S}_z|0, 0\rangle = \frac{\hbar}{2}|1, 0\rangle. \tag{3.3.26}$$

This allows us to write the time-dependent state vector as

$$|\psi(t)\rangle = \tilde{a}_1(t)|0, 0\rangle + \tilde{a}_2(t)|1, 0\rangle, \tag{3.3.27}$$

and let $\hat{H}$ of Eqs. (3.3.22) and (3.3.23) operate on it. We multiple from the left by the related bras and find with the assistance of

Eqs. (2.6.10) and (2.6.11)

$$i\hbar\frac{d\tilde{a}_1(t)}{dt} = -\frac{3}{4}A\hbar^2\tilde{a}_1(t) + \frac{\hbar}{2}\Omega_0\tilde{a}_2(t)\cos\omega t, \qquad (3.3.28)$$

$$i\hbar\frac{d\tilde{a}_2(t)}{dt} = \frac{1}{4}A\hbar^2\tilde{a}_2(t) + \frac{\hbar}{2}\Omega_0\tilde{a}_1(t)\cos\omega t. \qquad (3.3.29)$$

These last two equations simplify when, with the help of Eqs. (2.6.10) and (2.6.11), we set

$$\tilde{a}_1(t) = e^{i3A\hbar t/4}A_1(t), \qquad (3.3.30)$$

$$\hat{a}_2(t) = e^{-iA\hbar t/4}A_2(t). \qquad (3.3.31)$$

The results are

$$\frac{dA_1(t)}{dt} = -i\frac{\Omega_0}{2}e^{-iA\hbar t}A_2(t)\cos\omega t, \qquad (3.3.32)$$

$$\frac{dA_2(t)}{dt} = -i\frac{\Omega_0}{2}e^{+iA\hbar t}A_1(t)\cos\omega t. \qquad (3.3.33)$$

We now make the Rotating Wave Approximation (RWA), which we soon justify, and let $a_1(t) = A_1(t)$ and $a_2(t) = -A_2(t)$. We find

$$\frac{da_1(t)}{dt} = i\frac{\Omega_0}{4}e^{+i(\omega-A\hbar)t}a_2(t), \qquad (3.3.34)$$

$$\frac{da_2(t)}{dt} = i\frac{\Omega_0}{4}e^{-i(\omega-A\hbar)t}a_1(t). \qquad (3.3.35)$$

We see that Eqs. (3.3.34) and (3.3.35) are Eqs. (3.2.5) and (3.2.6) with several changes of variable and parameter names. In particular, we have $\Omega_0/2$ in place of $\tilde{\Omega}$. The initial conditions are

$$a_1(t = 0) = 1, \qquad (3.3.36)$$

$$a_2(t = 0) = 0. \qquad (3.3.37)$$

All of this leads to the occupation probabilities of Eqs. (3.2.36) and (3.2.37) and we identify $p_1(t)$ with that of $|0,0\rangle$ and $p_2(t)$ with that of $|1,0\rangle$.

We introduce the experimental data for $p_2(t)$ in Fig. 3.15 (Olmschenk *et al.*, 2007, Fig. 7b). The probability is seen to vary from zero to nearly one and back, so we assume we are at resonance. The second observation is that the curve is reasonably smooth and lacks obvious high frequency components. The microwave frequency is 12.6428 GHz. At resonance, Eq. (3.2.37) becomes in the present notation

$$p_2(t) = [\sin(\Omega_0 t/4)]^2. \tag{3.3.38}$$

This reaches 1.0 when $t = 6$ μs and we deduce

$$\Omega_0 = 2\pi/6 \times 10^{-6} = 1.05 \times 10^6 \,\text{Hz}. \tag{3.3.39}$$

At resonance the γ of Eq. (3.3.21) is

$$\gamma = 2(2\pi)12.643 \times 10^9/1.05 \times 10^6 = 1.51 \times 10^5, \tag{3.3.40}$$

and we have $\gamma \gg 1$. This and our assumption of resonance lead us to believe that the RWA will suffice here. We are building on our numerical results of Sec. 3.3.1 and the App. 5 finding that the high frequency components go as $1/\gamma$, so they are negligible here.

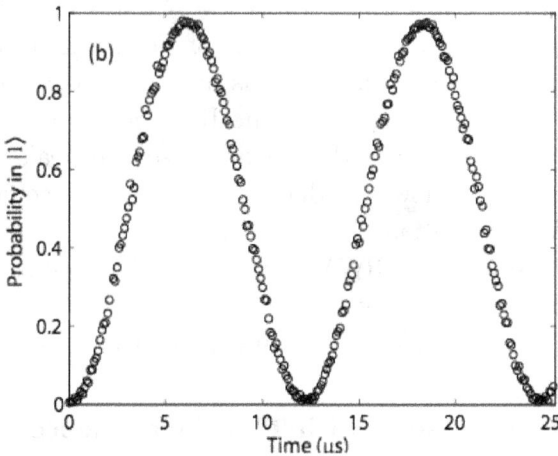

Figure 3.15. The probability of the occupation of the $|1, 0\rangle$ hyperfine level of ^{171}Yb$^+$ as measured by Olmschenk *et al.* (2007). Each point is the result of 1,000 measurements (Olmschenk et al., 2007, Fig. 7(b)). Reprinted figure with permission. Copyright 2007 by the American Physical Society.

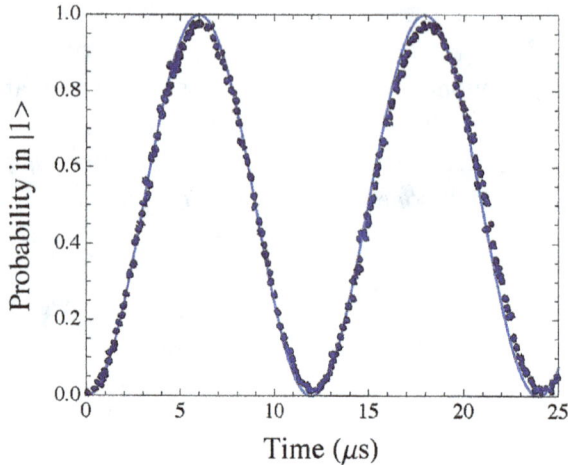

Figure 3.16. The data of Fig. 3.15 and the occupation probability for the excited state at resonance in the RWA with $\Omega_0 = 1.05 \times 10^6$ Hz. The system starts in the lower energy level of the hyperfine levels of $^{171}\mathrm{Yb}^+$. $p_2(t)$ is the solid blue line. The data points of Olmschenk *et al.* (2007, Fig. 7(b)) are traced onto the RWA results and the resulting figure is scanned.

In addition, $\Omega_0/2 \ll \omega_0 = 2\pi(12.643 \times 10^9)$, hence, the data's smooth curve.

We evaluate the RWA with the Ω_0 of Eq. (3.3.39) and plot the results in Fig. 3.16. The axes lengths are adjusted to compare the $p_2(t)$ with the data in Fig. 3.15. The RWA $p_2(t)$ overlays the first peak, but is very slightly to the left of the second peak.

This comparison shows the data are explained reasonably well by a simple RWA calculation.

We continue with the RWA in Sec. 3.4, where we learn how the time-dependent two-level system behaves when the potential energy for a constant field is added to the time-dependent coupling potential.

3.4. Two-Level System with Time Dependence and a Constant Field

Sections 3.1 and 3.2 developed the solution of the Time-Dependent Schrödinger Equation (TDSE) for a $\cos \omega t$ coupling between the two levels. It is common to also have a constant electric or magnetic field present and we now treat this case. We cast the new situation in terms of the Pauli spin matrices, $\{\sigma_i\}$, which are detailed in Sec. 1.3. The use of the Pauli spin matrices helps with the notation and gets us accustomed to their use in two-level problems. We work with the coupled equations and eventually get them into a form we encountered in Sec. 3.2.

The Hamiltonian $\hat{H}$ is taken to represent the interaction between a particle's magnetic moment and the magnetic fields. We assume a constant magnetic field B_z along the z-axis and an oscillating magnetic field $B_x \cos \omega t$ along the x-axis. The magnetic fields couple to a magnetic dipole moment

$$\hat{\vec{M}} = -g\mu_B \hat{\vec{J}}/\hbar, \tag{3.4.1}$$

where g is the Landé g-factor, μ_B is the Bohr magneton of the particle and $\hat{\vec{J}}$ is the total angular momentum operator for the particle. In Eq. (3.4.1) the minus sign is appropriate for negatively-charged particles such as electrons. The Landé g-factor is derived in Kroemer (1994, pp. 557–559) and is

$$g = 1 + \frac{j(j+1) + s(s+1) - l(l+1)}{2j(j+1)}. \tag{3.4.2}$$

In this equation, j is the quantum number for the total angular momentum, s is the spin quantum number and l is the orbital angular momentum quantum number. The Hamiltonian is

$$\hat{H} = -\hat{\vec{M}} \cdot \bar{B} = +\frac{g\mu_B}{\hbar}(\hat{J}_z B_z + \hat{J}_x B_x \cos \omega t). \tag{3.4.3}$$

We set $\bar{L} = 0$ and introduce the Pauli spin matrices through

$$\hat{\vec{J}} = \hat{\vec{S}} = \frac{\hbar}{2}\bar{\sigma}, \tag{3.4.4}$$

and the Hamiltonian becomes

$$\hat{H} = +\frac{g\mu_B}{2}(B_z \sigma_z + B_x \cos \omega t \sigma_x). \tag{3.4.5}$$

The wave function is now

$$\psi(t) = \begin{pmatrix} c_+(t) \\ c_-(t) \end{pmatrix}, \tag{3.4.6}$$

where $c_+(t)$ and $c_-(t)$ go with $m_j = +1/2$ and $m_j = -1/2$, respectively, and

$$|c_+(t)|^2 + |c_-(t)|^2 = 1. \tag{3.4.7}$$

When the Pauli spin matrices are written out, the TDSE becomes

$$i\hbar\frac{\partial}{\partial t}\begin{pmatrix} c_+(t) \\ c_-(t) \end{pmatrix} = \frac{g\mu_B}{2}\left[\begin{pmatrix} 1 & 0 \\ 0 & -1 \end{pmatrix} B_z + \begin{pmatrix} 0 & 1 \\ 1 & 0 \end{pmatrix} B_x \cos \omega t\right]\begin{pmatrix} c_+(t) \\ c_-(t) \end{pmatrix}. \tag{3.4.8}$$

We define the angular frequencies

$$\Omega_0 = g\mu_B B_z/\hbar, \tag{3.4.9}$$

$$\tilde{\Omega} = g\mu_B B_x/\hbar, \tag{3.4.10}$$

and write Eq. (3.4.8) as the coupled equations

$$i\frac{dc_+(t)}{dt} = \frac{\Omega_0}{2}c_+(t) + \frac{\tilde{\Omega}}{2}\cos \omega t c_-(t), \tag{3.4.11}$$

$$i\frac{dc_-(t)}{dt} = -\frac{\Omega_0}{2}c_-(t) + \frac{\tilde{\Omega}}{2}\cos \omega t c_+(t). \tag{3.4.12}$$

We next transform the last two equations in order to remove the Ω_0 terms,

$$\kappa_+(t) = c_+(t)e^{+i\Omega_0 t/2}, \tag{3.4.13}$$

$$\kappa_-(t) = c_-(t)e^{-i\Omega_0 t/2}, \tag{3.4.14}$$

so,

$$\left[i\frac{d\kappa_+(t)}{dt}e^{-i\Omega_0 t/2} + \frac{\Omega_0}{2}\kappa_+(t)e^{-i\Omega_0 t/2}\right]$$

$$= \frac{\Omega_0}{2}\kappa_+(t)e^{-i\Omega_0 t/2} + \frac{\tilde{\Omega}}{2}\kappa_-(t)\cos \omega t e^{+i\Omega_0 t/2}, \tag{3.4.15}$$

$$\left[i\frac{d\kappa_-(t)}{dt}e^{+i\Omega_0 t/2} - \frac{\Omega_0}{2}\kappa_-(t)e^{+i\Omega_0 t/2}\right]$$

$$= -\frac{\Omega_0}{2}\kappa_-(t)e^{+i\Omega_0 t/2} + \frac{\tilde{\Omega}}{2}\kappa_+(t)\cos \omega t e^{-i\Omega_0 t/2}. \tag{3.4.16}$$

The second term on the left-hand side cancels the first term on the right-hand side in each of the last two equations. Thus, the Ω_0 terms are removed. We clear the exponential factors on the left-hand sides to find

$$i\frac{d\kappa_+(t)}{dt} = \frac{\tilde{\Omega}}{2}\cos\omega t\, e^{i\Omega_0 t}\kappa_-(t), \tag{3.4.17}$$

$$i\frac{d\kappa_-(t)}{dt} = \frac{\tilde{\Omega}}{2}\cos\omega t\, e^{-i\Omega_0 t}\kappa_+(t). \tag{3.4.18}$$

We are almost there! The replacement of $\cos\omega t$ with

$$\cos\omega t = \frac{1}{2}(e^{i\omega t} + e^{-i\omega t}), \tag{3.4.19}$$

let us invoke the Rotating Wave Approximation. We are led to

$$i\frac{d\kappa_+(t)}{dt} = \frac{\tilde{\Omega}}{4}e^{i(\Omega_0-\omega)t}\kappa_-(t), \tag{3.4.20}$$

$$i\frac{d\kappa_-(t)}{dt} = \frac{\tilde{\Omega}}{4}e^{-i(\Omega_0-\omega)t}\kappa_+(t). \tag{3.4.21}$$

We are quite close to Eqs. (3.2.5) and (3.2.6). In fact, if we introduce a phase factor through,

$$\kappa_+ = i\kappa'_+, \tag{3.4.22}$$

$$\kappa_- = -i\kappa'_-, \tag{3.4.23}$$

then the coupled equations for $\kappa'_\pm$ have the same form as Eqs. (3.2.5) and (3.2.6). Thus, the solutions to Eqs. (3.4.20) and (3.4.21) are obtained by the method of Sec. 3.2. The above equations are useful for the discussion of magnetic resonance, a topic we take up in Chap. 4. We remark that in Sec. 3.2 an electric field was used, while here we treat a magnetic field. This leads to the extra factor of 2 that appears in Eqs. (3.4.20) and (3.4.21).

In the next section we take our two-level system into the Interaction Picture.

3.5. Two-Level Systems in the Interaction Picture

We now work in the Interaction Picture, also called the Dirac Picture, since it is useful for probing time-dependent problems such as those of this chapter. Our presentation follows Sakurai and Napolitano (2011, Sec. 5.5), but the notation is changed to agree with the earlier parts of this chapter. We use kets and find the time development of the state ket for the Hamiltonian of Eq. (3.5.1). We then express the Hamiltonian in terms of bras and kets and return to the two-level problem treated in Secs. 3.2 and 3.3.2. The resulting equations are simplified until equations analogous to Eqs. (3.2.5) and (3.2.6) emerge.

We start in the Schrödinger Picture with

$$\hat{H} = \hat{H}_0 + \hat{V}(t), \tag{3.5.1}$$

and a complete set of orthonormal eigenkets for $\hat{H}_0$

$$\hat{H}_0|\phi_n\rangle = H_n|\phi_n\rangle. \tag{3.5.2}$$

The Schrödinger state ket at time $t = 0$ is expressed with the subscript S

$$|\alpha, t_0; t = 0\rangle_S = \sum_n c_n(0)|\phi_n\rangle. \tag{3.5.3}$$

Then at time $t > 0$, the state ket evolves according to

$$|\alpha, t_0; t\rangle_S = \sum_n c_n(t)e^{-iH_nt/\hbar}|\phi_n\rangle. \tag{3.5.4}$$

The Interaction Picture allows us to concentrate on $\hat{V}(t)$. We first define the state ket in the Interaction Picture with the subscript I by

$$|\alpha, t_0; t\rangle_I = e^{i\hat{H}_0t/\hbar}|\alpha, t_0; t\rangle_S, \tag{3.5.5}$$

and we note at $t = 0$ the state kets in the two pictures are equal

$$|\alpha, t_0; 0\rangle_I = |\alpha, t_0; 0\rangle_S. \tag{3.5.6}$$

The next step is to derive the equivalent of the Time-Dependent Schrödinger Equation (TDSE). We start by taking the time derivative of Eq. (3.5.5)

$$i\hbar\frac{\partial}{\partial t}|\alpha, t_0; t\rangle_I = i\hbar\frac{\partial}{\partial t}(e^{i\hat{H}_0 t/\hbar}|\alpha, t_0; t\rangle_S)$$

$$= -\hat{H}_0 e^{i\hat{H}_0 t/\hbar}|\alpha, t_0; t\rangle_S + e^{i\hat{H}_0 t/\hbar}i\hbar\frac{\partial}{\partial t}|\alpha, t_0; t\rangle_S.$$

$$(3.5.7)$$

We recognize that the second term on the right-hand side is part of the TDSE, so

$$i\hbar\frac{\partial}{\partial t}|\alpha, t_0; t\rangle_I = -\hat{H}_0 e^{i\hat{H}_0 t/\hbar}|\alpha, t_0; t\rangle_S + e^{i\hat{H}_0 t/\hbar}(\hat{H}_0 + \hat{V})|\alpha, t_0; t\rangle_S.$$

$$(3.5.8)$$

The first term on the right-hand side contains the Interaction Picture state ket of Eq. (3.5.5) and the second term does also, since $\hat{H}_0$ commutes with the $\hat{H}_0$ in the exponential. We are led to

$$i\hbar\frac{\partial}{\partial t}|\alpha, t_0; t\rangle_I = -\hat{H}_0|\alpha, t_0; t\rangle_I + \hat{H}_0|\alpha, t_0; t\rangle_I$$

$$+e^{i\hat{H}_0 t/\hbar}\hat{V}e^{-i\hat{H}_0 t/\hbar}e^{i\hat{H}_0 t/\hbar}|\alpha, t_0; t\rangle_S. \qquad (3.5.9)$$

The first two terms on the right cancel, we use Eq. (3.5.5) to find

$$i\hbar\frac{\partial}{\partial t}|\alpha, t_0; t\rangle_I = e^{i\hat{H}_0 t/\hbar}\hat{V}e^{-i\hat{H}_0 t/\hbar}|\alpha, t_0; t\rangle_I \equiv \hat{V}_I(t)|\alpha, t_0; t\rangle_I,$$

$$(3.5.10)$$

and we have defined the potential operator in the Interaction Picture $\hat{V}_I(t)$. Equation (3.5.10) provides the Interaction Picture equation for the time development of the state ket.

We return to our two-level system and we express both $\hat{H}_0$ and the potential operator that couples levels 1 and 2 in terms of bras and kets. Eventually, we bring in Eq. (3.5.10). We start with

$$\hat{H}_0 = H_1|\phi_1\rangle\langle\phi_1| + H_2|\phi_2\rangle\langle\phi_2|. \qquad (3.5.11)$$

Here, H_1 and H_2 are the eigenenergies of $\hat{H}_0$. We multiply from the right with $|\phi_1\rangle$

$$\hat{H}_0|\phi_1\rangle = H_1|\phi_1\rangle\langle\phi_1|\phi_1\rangle + H_2|\phi_2\rangle\langle\phi_2|\phi_1\rangle = H_1|\phi_1\rangle. \qquad (3.5.12)$$

The second equality follows from the assumed orthonormality of the $|\phi_i\rangle$ captured in Eq. (3.1.5). Similarly,

$$\hat{H}_0|\phi_2\rangle = H_2|\phi_2\rangle. \qquad (3.5.13)$$

Now we express the Schrödinger Picture $\hat{V}(t)$ as

$$\hat{V}(t) = -\gamma e^{i\omega t}|\phi_1\rangle\langle\phi_2| - \gamma e^{-i\omega t}|\phi_2\rangle\langle\phi_1|. \qquad (3.5.14)$$

This form is equivalent to the Rotating Wave Approximation (RWA) with γ specified below. Here ω is the angular frequency of our time-dependent classical electromagnetic field and we assume γ is real. We note this γ is not the γ of Sec. 3.3.2. Equation (3.5.14) takes $|\phi_1\rangle$ to $|\phi_2\rangle$

$$\hat{V}(t)|\phi_1\rangle = -\gamma e^{i\omega t}|\phi_1\rangle\langle\phi_2|\phi_1\rangle$$
$$- \gamma e^{-i\omega t}|\phi_2\rangle\langle\phi_1|\phi_1\rangle = -\gamma e^{-i\omega t}|\phi_2\rangle,$$
$$\qquad (3.5.15)$$

and $|\phi_2\rangle$ to $|\phi_1\rangle$

$$\hat{V}(t)|\phi_2\rangle = -\gamma e^{i\omega t}|\phi_1\rangle\langle\phi_2|\phi_2\rangle$$
$$- \gamma e^{-i\omega t}|\phi_2\rangle\langle\phi_1|\phi_2\rangle = -\gamma e^{+i\omega t}|\phi_1\rangle.$$
$$\qquad (3.5.16)$$

We note that this form of $\hat{V}(t)$ leads to

$$\langle\phi_1|\hat{V}(t)|\phi_1\rangle = -\gamma e^{-i\omega t}\langle\phi_1|\phi_2\rangle = 0, \qquad (3.5.17)$$

$$\langle\phi_2|\hat{V}(t)|\phi_2\rangle = -\gamma e^{+i\omega t}\langle\phi_2|\phi_1\rangle = 0. \qquad (3.5.18)$$

Finally, in the Interaction Picture, the off-diagonal potential operator matrix elements are

$$\langle\phi_i|\hat{V}_I(t)|\phi_j\rangle = \langle\phi_i|e^{i\hat{H}_0 t/\hbar}\hat{V}(t)e^{-i\hat{H}_0 t/\hbar}|\phi_j\rangle$$
$$= e^{iH_i t/\hbar}\langle\phi_i|\hat{V}(t)|\phi_j\rangle e^{-iH_j t/\hbar}. \qquad (3.5.19)$$

We continue with an expression for the state ket in the Interaction Picture,

$$|\alpha, t_0; t\rangle_I = d_1(t)|\phi_1\rangle + d_2(t)|\phi_2\rangle, \qquad (3.5.20)$$

and we need to solve for the $d_i(t)$. We return to Eq. (3.5.10), multiply from the left by $\langle\phi_1|$, and use Eq. (3.5.20) to find

$$i\hbar\left(\langle\phi_1|\phi_1\rangle\frac{d}{dt}d_1(t) + \langle\phi_1|\phi_2\rangle\frac{d}{dt}d_2(t)\right)$$
$$= \langle\phi_1|\hat{V}_I(t)|\phi_1\rangle d_1(t) + \langle\phi_1|\hat{V}_1(t)|\phi_2\rangle d_2(t). \qquad (3.5.21)$$

This is reduced through the orthonormality of the $|\phi_i\rangle$ and the above matrix elements to

$$i\hbar\frac{d}{dt}d_1(t) = \langle\phi_1|\hat{V}_I(t)|\phi_2\rangle d_2(t) = -\gamma e^{i\omega t}e^{i(H_1-H_2)t/\hbar}d_2(t).$$
$$(3.5.22)$$

Similarly, multiplication of Eq. (3.5.10) from the left by $\langle\phi_2|$ and the use of Eq. (3.5.20) yields

$$i\hbar\frac{d}{dt}d_2(t) = \langle\phi_2|\hat{V}_I(t)|\phi_1\rangle d_1(t) = -\gamma e^{-i\omega t}e^{i(H_2-H_1)t/\hbar}d_1(t).$$
$$(3.5.23)$$

We next set $\omega_i = H_i/\hbar$ and identify

$$\gamma = \mu E/2. \qquad (3.5.24)$$

We once again set

$$\delta\omega = \omega - \omega_0 = \omega - (\omega_2 - \omega_1). \qquad (3.5.25)$$

With these definitions and with Eq. (3.2.4) in mind, we rewrite Eqs. (3.5.22) and (3.5.23) as

$$\frac{d}{dt}d_1(t) = [i\mu E/(2\hbar)]e^{i\delta\omega t}d_2(t), \qquad (3.5.26)$$

$$\frac{d}{dt}d_2(t) = [i\mu E/(2\hbar)]e^{-i\delta\omega t}d_1(t). \qquad (3.5.27)$$

These equations have the identical form that Eqs. (3.2.5) and (3.2.6) do. Thus, the Interaction Picture and the coupling potential of Eq. (3.5.14) have led us to a set of coupled equations that we have solved in Sec. 3.2. We emphasize that the RWA has been built into this set of equations through Eq. (3.5.14). The exponentials in Eqs. (3.5.26) and (3.5.27) are removed by

$$d_1'(t) = d_1(t)$$
$$d_2'(t) = e^{i\delta\omega t}d_2(t), \tag{3.5.28}$$

so,

$$\frac{d}{dt}d_1'(t) = [i\mu E/(2\hbar)]d_2'(t), \tag{3.5.29}$$

$$\frac{d}{dt}d_2'(t) = [i\mu E/(2\hbar)]d_1'(t) + i\delta\omega d_2'(t). \tag{3.5.30}$$

Now we must to show how the $d_i(t)$ of Eq. (3.5.20) and the $c_i(t)$ of Eq. (3.5.4) are related. We start with Eq. (3.5.20) and use Eqs. (3.5.4) and (3.5.5)

$$
\begin{aligned}
|\alpha, t_0; t_1\rangle_I &= d_1(t)|\phi_1\rangle + d_2(t)|\phi_2\rangle = e^{i\hat{H}_0 t/\hbar}|\alpha, t_0; t\rangle_S \\
&= e^{i\hat{H}_0 t/\hbar}(c_1(t)e^{-iH_1 t/\hbar}|\phi_1\rangle + c_2(t)e^{-iH_2 t/\hbar}|\phi_2\rangle) \\
&= c_1(t)e^{-iH_1 t/\hbar}e^{+iH_1 t/\hbar}|\phi_1\rangle + c_2(t)e^{-iH_2 t/\hbar}e^{+iH_2 t/\hbar}|\phi_2\rangle \\
&= c_1(t)|\phi_1\rangle + c_2(t)|\phi_2\rangle. \tag{3.5.31}
\end{aligned}
$$

Hence,

$$d_i(t) = c_i(t). \tag{3.5.32}$$

Thus, the occupation probabilities we derived in Sec. 3.2 apply here. We return to the Interaction Picture in Chap. 5.

We continue in Sec. 3.6 with a two-level system that is solvable without the need for the RWA.

3.6. Two-Level Systems with a Rotating Field

We have seen in Secs. 3.1 to 3.4 that the coupled equations for the wave function with a $\cos\omega t$ coupling are solvable in closed form when the Rotating Wave Approximation (RWA) is made. The cosine term is linked to a linearly-polarized electromagnetic field, which may be viewed as a superposition of two circularly-polarized electromagnetic fields. In this section, we investigate the case of a circularly-polarized electromagnetic field. It is surprising that this case is solvable without the need for the RWA. There are parallels to the coupling potential operator, Eq. (3.5.14), treated in Sec. 3.5 in the Interaction Picture. However, in the present section we also include a constant magnetic field and we work within the Schrödinger Picture.

Hecht (2000, Chap. 58) treats the circular polarization by constructing the time-evolution operator $U(t,0)$. We now take an approach in line with the earlier sections of this chapter and follow Cohen-Tannoudji *et al.* (1977). They clearly set up the circularly-polarized case in their Complement F_{IV}. They start with a combination of a static and a rotating magnetic field and show how the classical magnetic moment changes with time. We commence with the quantum mechanical treatment and present the derivation in terms of a spin S = 1/2 system. The two levels are spin-up, denoted by $|+\rangle$, and spin-down denoted by $|-\rangle$, and these are defined with respect to the z-axis. Our notation follows that used earlier in this chapter.

As in our Eqs. (3.4.1) and (3.4.3), we take the Hamiltonian to be

$$\hat{H} = -\hat{\vec{M}} \cdot \vec{B}(t), \qquad (3.6.1)$$

with $\hat{\vec{J}} = \hat{\vec{S}}$. Then the present case is specified by

$$\hat{H} = \frac{g\mu_B}{\hbar}(\hat{S}_z B_0 + \hat{S}_x B_1 \cos\omega t + \hat{S}_y B_1 \sin\omega t). \qquad (3.6.2)$$

The component of the electromagnetic field in the x-y plane is of constant magnitude and rotates with angular frequency ω. We set

$$\Omega_0 = g\mu_B B_0/\hbar, \qquad (3.6.3)$$

$$\tilde{\Omega} = g\mu_B B_1/\hbar, \qquad (3.6.4)$$

so

$$\hat{H} = \Omega_0 \hat{S}_z + \tilde{\Omega}(\hat{S}_x \cos\omega t + \hat{S}_y \sin\omega t). \qquad (3.6.5)$$

We note when $B_1 = 0$, the $|+\rangle$ and the $|-\rangle$ are the eigenkets of $\Omega_0 \hat{S}_z$ with the eigenenergies of $\frac{\hbar}{2}\Omega_0$ and $-\frac{\hbar}{2}\Omega_0$, respectively. Thus, B_0 sets the condition for resonance.

We now rewrite the Hamiltonian in terms of the Pauli matrices of Sec. 1.3 and we use the $\{|+\rangle, |-\rangle\}$,

$$\hat{H} = \frac{\hbar\Omega_0}{2}\begin{pmatrix} 1 & 0 \\ 0 & -1 \end{pmatrix} + \frac{\hbar\tilde{\Omega}}{2}\left[\cos\omega t \begin{pmatrix} 0 & 1 \\ 1 & 0 \end{pmatrix} + \sin\omega t \begin{pmatrix} 0 & -i \\ i & 0 \end{pmatrix}\right]. \quad (3.6.6)$$

Now

$$\cos\omega t \pm i\sin\omega t = \frac{1}{2}(e^{i\omega t} + e^{-i\omega t}) \pm \frac{1}{2}(e^{i\omega t} - e^{-i\omega t}). \quad (3.6.7)$$

The plus sign yields $e^{i\omega t}$ and the minus signs gives $e^{-i\omega t}$. These let us rewrite the Hamiltonian in Eq. (3.6.6) as

$$\hat{H} = \frac{\hbar}{2}\begin{pmatrix} \Omega_0 & \tilde{\Omega}e^{-i\omega t} \\ \tilde{\Omega}e^{i\omega t} & -\Omega_0 \end{pmatrix}. \quad (3.6.8)$$

The ket for our two-level system is

$$|\psi(t)\rangle = a_+(t)|+\rangle + a_-(t)|-\rangle = \begin{pmatrix} a_+(t) \\ a_-(t) \end{pmatrix}. \quad (3.6.9)$$

We apply the Hamiltonian of Eq. (3.6.8) to this state and multiply from the left by $\langle+|$. The assumed orthonormality of the basis kets leads to Eq. (3.6.10). We then start again and multiply from the left by $\langle-|$. This provides Eq. (3.6.11) and the Time-Dependent Schrödinger Equation becomes the coupled equations

$$i\hbar\frac{da_+(t)}{dt} = \frac{\hbar}{2}\Omega_0 a_+(t) + \frac{\hbar}{2}\tilde{\Omega}e^{-i\omega t}a_-(t), \quad (3.6.10)$$

$$i\hbar\frac{da_-(t)}{dt} = \frac{\hbar}{2}\tilde{\Omega}e^{i\omega t}a_+(t) - \frac{\hbar}{2}\Omega_0 a_-(t). \quad (3.6.11)$$

The following transformation removes the time dependence in the coefficients of the last two equations and is a transformation to a rotating frame,

$$b_+(t) = e^{i\omega t/2}a_+(t), \quad (3.6.12)$$

$$b_-(t) = e^{-i\omega t/2}a_-(t). \quad (3.6.13)$$

Thus, we place the $b_{\pm}(t)$ into Eqs. (3.6.10) and (3.6.11) and divide out $\hbar$. We find

$$ie^{-i\omega t/2}\frac{db_+(t)}{dt} + \frac{\omega}{2}e^{-i\omega t/2}b_+(t)$$

$$= \frac{\Omega_0}{2}e^{-i\omega t/2}b_+(t) + \frac{\tilde{\Omega}}{2}e^{-i\omega t/2}b_-(t), \qquad (3.6.14)$$

$$ie^{i\omega t/2}\frac{db_-(t)}{dt} - \frac{\omega}{2}e^{i\omega t/2}b_-(t)$$

$$= \frac{\tilde{\Omega}}{2}e^{i\omega t/2}b_+(t) - \frac{\Omega_0}{2}e^{i\omega t/2}b_-(t). \qquad (3.6.15)$$

As noted above, the difference in the eigenenergies of the two levels with $B_1 = 0$ is $\hbar\Omega_0$. So, we write

$$\delta\omega = \omega - \Omega_0, \qquad (3.6.16)$$

with $\delta\omega = 0$ for resonance. We also cancel the exponentials and see the coupled equations

$$i\frac{db_+(t)}{dt} = -\frac{\delta\omega}{2}b_+(t) + \frac{\tilde{\Omega}}{2}b_-(t), \qquad (3.6.17)$$

$$i\frac{db_-(t)}{dt} = \frac{\tilde{\Omega}}{2}b_+(t) + \frac{\delta\omega}{2}b_-(t). \qquad (3.6.18)$$

And we note that the coefficients on the right-hand sides of the coupled equations are now independent of time! We return to the realm of Chap. 2. In the present chapter, the higher energy state is $|+\rangle$, whereas in Chap. 2 it is $|\phi_2\rangle$.

We next assume

$$b_{\pm}(t) = e^{-i\Omega t}b_{\pm}, \qquad (3.6.19)$$

and find

$$i(-i\Omega)e^{-i\Omega t}b_+ = -\frac{\delta\omega}{2}e^{-i\Omega t}b_+ + \frac{\tilde{\Omega}}{2}e^{-i\Omega t}b_-, \qquad (3.6.20)$$

$$i(-i\Omega)e^{-i\Omega t}b_- = \frac{\tilde{\Omega}}{2}e^{-i\Omega t}b_+ + \frac{\delta\omega}{2}e^{-i\Omega t}b_-. \qquad (3.6.21)$$

The exponential terms cancel and we rewrite these coupled equations in matrix form

$$\Omega \begin{pmatrix} b_+ \\ b_- \end{pmatrix} = \begin{pmatrix} -\dfrac{\delta\omega}{2} & \dfrac{\tilde{\Omega}}{2} \\[2mm] \dfrac{\tilde{\Omega}}{2} & \dfrac{\delta\omega}{2} \end{pmatrix} \begin{pmatrix} b_+ \\ b_- \end{pmatrix}. \tag{3.6.22}$$

The eigenvalues Ω follow from

$$\left(-\frac{\delta\omega}{2} - \Omega \right) \left(\frac{\delta\omega}{2} - \Omega \right) - \frac{\tilde{\Omega}^2}{4} = -\frac{(\delta\omega)^2}{4} + \Omega^2 - \frac{\tilde{\Omega}^2}{4} = 0, \tag{3.6.23}$$

which leads to

$$\Omega_\pm = \pm\frac{1}{2}\sqrt{\tilde{\Omega}^2 + (\delta\omega)^2}. \tag{3.6.24}$$

We use these eigenvalues to find the eigenkets, in order to evaluate the occupation probability of the system being in $|+\rangle$ or in $|-\rangle$, when a measurement is made at time t. We start with Ω_+ and introduce b_{1+} and b_{2+}

$$\begin{pmatrix} -\dfrac{\delta\omega}{2} & \dfrac{\tilde{\Omega}}{2} \\[2mm] \dfrac{\tilde{\Omega}}{2} & \dfrac{\delta\omega}{2} \end{pmatrix} \begin{pmatrix} b_{1+} \\ b_{2+} \end{pmatrix} = \frac{1}{2}\sqrt{\tilde{\Omega}^2 + (\delta\omega)^2} \begin{pmatrix} b_{1+} \\ b_{2+} \end{pmatrix}. \tag{3.6.25}$$

Then, the first equation yields

$$-\frac{\delta\omega}{2}b_{1+} + \frac{\tilde{\Omega}}{2}b_{2+} = \frac{1}{2}\sqrt{\tilde{\Omega}^2 + (\delta\omega)^2}\,b_{1+}, \tag{3.6.26}$$

which is

$$b_{1+} = \tilde{\Omega}b_{2+}\left(\sqrt{\tilde{\Omega}^2 + (\delta\omega)^2} + \delta\omega \right)^{-1}. \tag{3.6.27}$$

Normalization of the eigenket requires

$$|b_{1+}|^2 + |b_{2+}|^2 = 1, \tag{3.6.28}$$

and this becomes

$$\frac{\tilde{\Omega}^2 |b_{2+}|^2}{\left(\sqrt{\tilde{\Omega}^2 + (\delta\omega)^2} + \delta\omega\right)^2} + |b_{2+}|^2 = 1. \tag{3.6.29}$$

Thus,

$$|b_{2+}|^2 = \frac{\left(\sqrt{\tilde{\Omega}^2 + (\delta\omega)^2} + \delta\omega\right)^2}{\left(\sqrt{\tilde{\Omega}^2 + (\delta\omega)^2} + \delta\omega\right)^2 + \tilde{\Omega}^2}, \tag{3.6.30}$$

and we take the positive root

$$b_{2+} = \frac{\left(\sqrt{\tilde{\Omega}^2 + (\delta\omega)^2} + \delta\omega\right)}{\left[\left(\sqrt{\tilde{\Omega}^2 + (\delta\omega)^2} + \delta\omega\right)^2 + \tilde{\Omega}^2\right]^{1/2}}. \tag{3.6.31}$$

With Eq. (3.6.27), this leads to

$$b_{1+} = \frac{\tilde{\Omega}}{\left[\left(\sqrt{\tilde{\Omega}^2 + (\delta\omega)^2} + \delta\omega\right)^2 + \tilde{\Omega}^2\right]^{1/2}}. \tag{3.6.32}$$

The second equation of Eq. (3.6.25) leads to the same results.

We repeat the calculation with Ω_- of Eq. (3.6.24) in order to find the related b_{1-} and b_{2-}. With the second equation of Eq. (3.6.22), this leads to

$$\frac{\tilde{\Omega}}{2} b_{1-} + \frac{\delta\omega}{2} b_{2-} = -\frac{1}{2}\sqrt{\tilde{\Omega}^2 + (\delta\omega)^2} b_{2-}, \tag{3.6.33}$$

and

$$b_{1-} = -\left(\sqrt{\tilde{\Omega}^2 + (\delta\omega)^2} + \delta\omega\right)(\tilde{\Omega})^{-1} b_{2-}. \tag{3.6.34}$$

We use the normalization condition to eliminate b_{1-} and find

$$b_{2-} = \frac{\tilde{\Omega}}{\left[\left(\sqrt{\tilde{\Omega}^2 + (\delta\omega)^2} + \delta\omega\right)^2 + \tilde{\Omega}^2\right]^{1/2}}, \tag{3.6.35}$$

and

$$b_{1-} = -\frac{\left(\sqrt{\tilde{\Omega}^2 + (\delta\omega)^2} + \delta\omega\right)}{\left[\left(\sqrt{\tilde{\Omega}^2 + (\delta\omega)^2} + \delta\omega\right)^2 + \tilde{\Omega}^2\right]^{1/2}}. \tag{3.6.36}$$

As in Chap. 2, we set the $b_{i\pm}$ to cosine or sine of an angle θ. Here we choose the $\sin\theta$ to go with the $b_{i\pm}$ that vanishes when $\tilde{\Omega} \to 0$, thus,

$$b_{1+} = \sin\theta, \tag{3.6.37}$$

$$b_{2+} = \cos\theta, \tag{3.6.38}$$

$$b_{1-} = -\cos\theta, \tag{3.6.39}$$

$$b_{2-} = \sin\theta. \tag{3.6.40}$$

We proceed to develop expressions for the occupation probabilities $p_\pm(t)$. It helps to recall we started with

$$|\psi(t)\rangle = a_+(t)|+\rangle + a_-(t)|-\rangle, \tag{3.6.9}$$

and then went to the $b_i(t)$ with Eqs. (3.6.12) and (3.6.13). We then used Eq. (3.6.19) to separate the time dependence and find the eigenkets. So, to within a phase factor, we write

$$\begin{pmatrix} |b_+(t)\rangle \\ |b_-(t)\rangle \end{pmatrix} = c_+ e^{-i\Omega_+ t} \begin{pmatrix} \sin\theta|+\rangle \\ \cos\theta|-\rangle \end{pmatrix} + c_- e^{-i\Omega_- t} \begin{pmatrix} -\cos\theta|+\rangle \\ \sin\theta|-\rangle \end{pmatrix}, \tag{3.6.41}$$

and we are using Eqs. (3.6.24) and (3.6.37) to (3.6.40). The $c_\pm$ are set by our choice of the initial conditions at $t = 0$. We start the system in the higher energy level state $|+\rangle$ at time $t = 0$, and Eq. (3.6.41) becomes the two equations

$$1 = \sin\theta c_+ - \cos\theta c_-, \tag{3.6.42}$$

$$0 = \cos\theta c_+ + \sin\theta c_-. \tag{3.6.43}$$

These two equations are solved by

$$c_+ = \sin\theta, \tag{3.6.44}$$

$$c_- = -\cos\theta. \tag{3.6.45}$$

Hence, we write Eq. (3.6.41) as

$$\begin{pmatrix} |b_+(t)\rangle \\ |b_-(t)\rangle \end{pmatrix} = \sin\theta e^{-i\Omega_+ t} \begin{pmatrix} \sin\theta|+\rangle \\ \cos\theta|-\rangle \end{pmatrix} - \cos\theta e^{-i\Omega_- t} \begin{pmatrix} -\cos\theta|+\rangle \\ \sin\theta|-\rangle \end{pmatrix}.$$

$$(3.6.46)$$

We are now ready to calculate the occupation probabilities. First, we evaluate the amplitude b_- that at time t the system is in level $|-\rangle$. This tells us the magnetic fields have induced at least one transition from our initial level $|+\rangle$ to $|-\rangle$. Equation (3.6.24) says $\Omega_- = -\Omega_+$, so we use Eq. (3.6.46) to write

$$b_-(t) = \sin\theta\cos\theta\Big(e^{-i\Omega_+ t} - e^{+i\Omega_+ t}\Big)$$

$$= -2i\sin\theta\cos\theta\sin\Big(\sqrt{\tilde{\Omega}^2 + (\delta\omega)^2}t/2\Big). \qquad (3.6.47)$$

The occupation probability follows

$$p_-(t) = b_-(t)b_-^*(t) = 4(\sin\theta\cos\theta)^2\sin^2\Big(\sqrt{\tilde{\Omega}^2 + (\delta\omega)^2}t/2\Big),$$

$$(3.6.48)$$

and we rewrite this as

$$p_-(t) = (\sin 2\theta)^2\sin^2\Big(\sqrt{\tilde{\Omega}^2 + (\delta\omega)^2}t/2\Big). \qquad (3.6.49)$$

In fact, we proceed in parallel to Sec. 2.2 and note that the present $\sin\theta$ and $\cos\theta$ lead, through Eqs. (3.6.32) and (3.6.31), respectively, to

$$\tan 2\theta = \sin 2\theta/\cos 2\theta = \frac{2\sin\theta\cos\theta}{(\cos\theta)^2 - (\sin\theta)^2} = \frac{\tilde{\Omega}}{\delta\omega}. \qquad (3.6.50)$$

This last expression allows us to rewrite Eq. (3.6.49) in the form

$$p_-(t) = \frac{\tilde{\Omega}^2}{[\tilde{\Omega}^2 + (\delta\omega)^2]}\sin^2\Big(\sqrt{\tilde{\Omega}^2 + (\delta\omega)^2}t/2\Big), \qquad (3.6.51)$$

and, indeed, this closely resembles Eq. (2.3.11). Thus, we expect Rabi oscillations for the present case of a circularly-polarized electromagnetic field.

As a check, we compute the occupation probability $p_+(t)$ for our initial condition of starting in $|+\rangle$ at $t = 0$. Equation (3.6.46) tells us

$$b_+(t) = \sin\theta\sin\theta e^{-i\Omega_+ t} + \cos\theta\cos\theta e^{+i\Omega_+ t}. \qquad (3.6.52)$$

Thus,

$$p_+(t) = b_+(t)b_+^*(t) = (\sin\theta)^4 + (\cos\theta)^4$$
$$+2(\sin\theta\cos\theta)^2 \cos\left(\sqrt{\tilde{\Omega}^2 + (\delta\omega)^2}t\right), \qquad (3.6.53)$$

and this is simplified using $1 = [(\sin\theta)^2 + (\cos\theta)^2]^2$ and the relation between the cosine of an angle and the sine of half the angle,

$$\cos\left(\sqrt{\tilde{\Omega}^2 + (\delta\omega)^2}t\right) = 1 - 2\left\{\sin\sqrt{\tilde{\Omega}^2 + (\delta\omega)^2}t/2\right\}^2. \qquad (3.6.54)$$

The result is

$$p_+(t) = 1 - (\sin 2\theta)^2 \sin^2\left(\sqrt{\tilde{\Omega}^2 + (\delta\omega)^2}t/2\right) = 1 - p_-(t). \quad (3.6.55)$$

This result is reassuring as it shows the sum of the occupation probabilities is always 1. The occupation probabilities do depend on the constant magnetic field B_0. It is buried in $\delta\omega$ through Eqs. (3.6.16) and (3.6.3) and it helps determine the energy and time scales of the Rabi oscillations.

We note that the solutions for circular polarization, and, hence, the occupation probabilities, involve only the angular frequency $\sqrt{\tilde{\Omega}^2 + (\delta\omega)^2}/2$. In fact, the occupation probabilities are those for the Rotating Wave Approximation (RWA) for the linear polarization case. We do not find what occurs for linear polarization when one goes beyond the RWA and discovers higher frequencies as part of the occupation probabilities. This occurrence is discussed in Sec. 3.3.2 and further explained in Cohen-Tannoudji *et al.* (1977, Complement B_{XIII}). The key is that circular polarization permits the transformation of Eqs. (3.6.12) and (3.6.13) and the resulting removal of time-dependent coefficients in the differential equations. Experimental data do show the absence of the higher frequencies when circular polarization is used (London *et al.*, 2014).

Chapter 4 continues with examples of two-level systems such as atoms, the ammonia maser, magnetic resonance phenomena and two-level defects in diamond. The results we just derived are used in the discussion of magnetic resonance.

References

D. Bonacci, "Rabi spectra — a simple tool for analyzing the limitations of RWA in modelling of the selective population transfer in many-level quantum systems", arXiv: quant-ph/0309126v4, 7 May 2004.

C. Cohen-Tannoudji, B. Diu and F. Laloë, *Quantum Mechanics*, Vols. 1 and 2 (John Wiley and Sons, New York, 1977).

C. Cohen-Tannoudji, J. Dupont-Roc, and G. Grynberg, *Atom-Photon Interactions: Basic Processes and Applications* (John Wiley and Sons, New York, 1992).

H. A. Enge, *Introduction to Nuclear Physics* (Addison-Wesley Publishing, Reading, MA, 1966).

M. Fox, *Quantum Optics: An Introduction* (Oxford University Press, Oxford, 2006), reprinted 2007.

G. D. Fuchs, V. V. Dobrovitski, D. M. Toyli, F. J. Heremans, and D. D. Awschalom, "Gigahertz Dynamics of a Strongly Driven Single Quantum Spin", *Science* **326**, 1520–1522 (2009).

K. T. Hecht, *Quantum Mechanics* (Springer, New York, 2000).

W. Ketterle, Physics 8.421, MIT Open Course Ware, Spring 2014, Lecture 13.

H. Kroemer, *Quantum Mechanics: For Engineering, Materials Science, and Applied Physics* (Prentice Hall, Upper Saddle River, N. J., 1994).

P. London, P. Balasubramanian, B. Naydenov. L. P. McGuinness, and F. Jelezko, "Strong driving of a single spin using arbitrarily polarized fields", *Phys. Rev.* A**90**, 012302 (2014).

Mathematica = Wolfram Research, Inc, Mathematica, Version 9.0.0.0, Champaign, IL (2012).

S. Olmschenk, K. C. Younge, D. L. Moehring, D. N. Matsukevich, P. Maunz, and C. Monroe, "Manipulation and detection of a trapped Yb^+ hyperfine qubit", *Phys. Rev.* A**76**, 052314 (2007).

H. Perrin, "Light forces", Les Houches lectures on laser cooling and trapping, 16–26 September 2014.

C. Roos, "Quantum information processing with trapped ions", in *Fundamental Physics in Particle Traps*, W. Quint and M. Vogel, eds. (Springer-Verlag, Berlin, 2014), Chap. 8.

J. J. Sakurai and J. Napolitano, *Modern Quantum Mechanics*, 2nd ed. (Addison-Wesley, Boston, 2011).

E. J. Sie, C. H. Lui, Y.-H. Lee, L. Fu, J. Kong, and N. Gedik, "Large, valley-exclusive Bloch–Siegert shift in monolayer WS$_2$", *Science*, **355**, 1066–1069 (2017).

B. W. Shore, *The Theory of Coherent Atomic Excitation*, Vol. 1 Simple Atoms and Fields (John Wiley and Sons, New York, 1990).

Chapter 4

Applications and Examples of Two-Level Systems

We now apply the formalism and machinery of Chap. 3 to several examples of two-level systems that evolve with time. Chapter 3 shows how Rabi oscillations (or Rabi flopping) arise and includes a physical example. Experimental demonstrations of Rabi oscillations are quite involved. The interested reader should consult the cited references to supplement the material that follows. We need to prepare a two-level system in a known state and we need a method to measure which state the system is in as a function of time. These requirements prove to be challenging and how they are satisfied varies with the two-level system. Historically, the first prediction and first observation of such oscillations were made by Rabi and involved radio frequencies (Rabi, 1937; Rabi *et al.*, 1938a). This work led to the use of magnetic resonance to probe materials. Here, a frequency or a field is adjusted to produce data analogous to Fig. 3.5. Further work with microwaves led to the demonstration of a MASER (Microwave Amplification by Stimulated Emission of Radiation) (Gordon *et al.*, 1954).

Eventually, experimental techniques were developed that permitted the observation of Rabi oscillations in the near visible, and this is where we start in Sec. 4.1. Ammonia masers are considered in Sec. 4.2. Bertolotti (2005, Chaps. 8 and 9) provides a brief history of masers, while Hecht (2005) details how masers led to atomic clocks and Major (1998) tells how atomic clocks work. Section 4.3 continues with a treatment of magnetic resonance and shows how two-level systems are routinely exploited to learn about materials. Section 4.4 explores Rabi oscillations of two-level systems based on defects in diamonds and includes a comparison between calculations and data. Finally, Sec. 4.5 has Ramble 4 on the Pulse Area Theorem that plays a role in Sec. 4.1.

As we delve into experimental data, three points are worth emphasizing. First, it is useful to consider the expected time dependences involved in the experimental data. The sample times and intervals need to be fine enough to catch the expected physical mechanisms in action. This is especially so when the data have an on/off nature. Second, while two-level signals are common and are generally attributed to a two-level system, a third level or an additional two- or more-level system may lurk nearby. It is often a challenge to decide whether the signal belongs to a two-level system or to a system with more levels. This topic is discussed in Chap. 6. Third, the nature of the on/off transition reveals much of the basic Physics. The behavior of an individual electron or spin generally drives the transition between on and off and then off and on. In many cases, we find quantum jumps, which are the focus of ongoing research (Muga *et al.*, 2008; Gleyzes *et al.*, 2007).

4.1. Rabi Oscillations Observed with Atomic Beams

Gibbs (1972; 1973) observed Rabi oscillations with a beam of rubidium atoms. A brief discussion of his work is found in Fox (2006, Sec. 9.5.3). The beam of ^{85}Rb and ^{87}Rb atoms emerged from an oven into a magnetic field of 7.45 tesla, where they were illuminated by a pulse from a Hg (mercury) laser. Rb atoms were continually entering and leaving the volume illuminated by the exciting photon pulses. Thus, the Rb atoms do not reach a steady state with the exciting laser photons. This avoids the limitation of two-level systems mentioned in Sec. 1.5 and at the end of Sec. 4.2. We will come back to this point when we discuss the replacement of the atomic beam by a solid.

The Hg laser provides a 6-nanosecond coherent optical pulse from a longer optical pulse. The shorter pulse was shaped to approximate a plane wave. A simplified energy level diagram is shown in Fig. 4.1

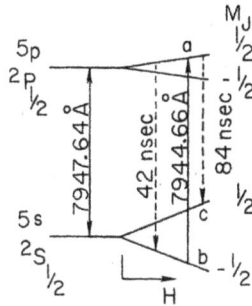

Figure 4.1. The energy level diagram for the Rb atoms, reproduced from Gibbs (1972, Fig. 2). Reprinted figure with permission. Copyright 1972 by the American Physical Society.

(Gibbs, 1972) with the two doublets. The constant magnetic field splits the Rb $5^2P_{1/2}$ and $5^2S_{1/2}$ lines into $m_J = \pm 1/2$ doublets by the Zeeman effect, as discussed in Townsend (2000, Chap. 11) and Konishi and Paffuti (2009, Chap. 15). The Hg laser pulse at 794.466 nm excites the electrons from the lower doublet's $m_J = -1/2$ state into the upper doublet's $m_J = +1/2$ state. This state then decays with a lifetime of 84 ns into the lower doublet's $m_J = +1/2$ state and with a lifetime of 42 ns into the lower doublet's $m_J = -1/2$ state. Gibbs estimates that for every 10^6 laser photons, one fluorescence photon results. The measurement aims to determine the population of the upper state after the pulse ends. Gibbs did this by counting the number of fluorescence photons of the 42-ns emission that resulted from the spontaneous emission. This fluorescence signal is integrated from 22 to 72 ns after the laser pulse arrives and is monitored as a function of the pulse area, which we explain next.

Our equations of Chap. 3 give the occupation probabilities as a function of time. We may define a pulse area that is proportional to an integral over time of the amplitude of the time-varying electromagnetic field. This connects our Eqs. (3.2.36) and (3.2.37) for the populations of the two levels with the data of Gibbs. The pulse area is considered further in Sec. 4.5. Figure 4.2 (Gibbs, 1973) shows his data for the integrated fluorescence signal versus the pulse area. The oscillation in the population of the upper $m_J = +1/2$ state with the pulse area is evident. A larger pulse area is equivalent

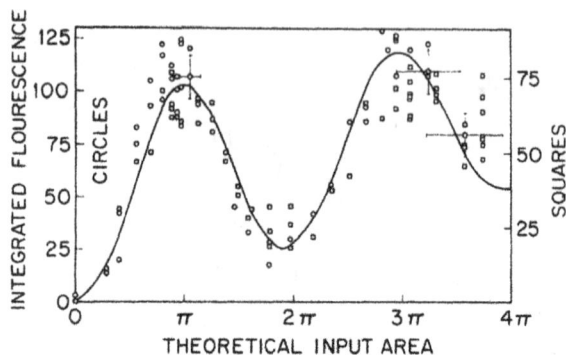

Figure 4.2. The measured integrated fluorescence versus the input pulse area (Gibbs, 1973, Fig. 14). The solid line is his calculated result and is explained in his paper. Reprinted figure with permission. Copyright 1973 by the American Physical Society.

to a longer time. And a pulse area of π ideally corresponds to a complete transfer of the population from the lower state to the upper state, which in turn leads to the maximum fluorescence signal. At resonance, Eq. (3.2.37) shows the π pulse is $\Omega t = \tilde{\Omega} t = \pi$. The minimum fluorescence signal occurs for a pulse area of 2π, so $\Omega t = 2\pi$ and $p_2 = 0$ per Eq. (3.2.37). Thus, the time variable is mapped to the pulse area and the Rabi oscillations appear as in Fig. 4.2.

Here we show that the equations of Sec. 3.2 describe the time dependence of the occupation probabilities and, hence, exhibit Rabi oscillations. Figure 4.1 explains how Gibbs studied a system that involves two states: the upper doublet's $m_J = +1/2$ state and the lower doublet's $m_J = -1/2$ state. Yet, at least one other level was present due to the 84 ns lifetime fluorescence and needed to be considered in his data analysis.

4.2. The Ammonia Maser

We continue with beams and now consider how a beam of ammonia molecules leads to maser action. A maser produces microwaves that are highly monoenergetic and coherent. Masers, like lasers, involve stimulated emission, which was introduced in Chap. 1. This emission requires a population of emitters, such as atoms or molecules in an upper energy level. If emission is to dominate absorption, then we need more of the atoms or molecules in the upper energy level than in the lower energy level. This constitutes a population inversion and was achieved by Gordon *et al.* (1954) with ammonia molecules. Bertolotti (2005) and Hecht (2005) provide extensive accounts of the developments and ideas that led to this first demonstration of maser action. Detailed discussions of the ammonia maser are found in Bransden and Joachain (2000, Sec. 16.3), Goswami (2003, Chap. 16) and Rigamonti and Carretta (2009, App. IX.1). In addition, McGervey (1983, Sec. 10.4) and Feynman *et al.* (1965, Chap. 9) provide enjoyable overviews of the ammonia maser. But how is the required population inversion obtained with ammonia molecules?

We return to the inversion doublets of ammonia, NH_3, that we discuss in Sec. 2.4. Each doublet leads to a two-level system. The lowest doublet is split by 9.84×10^{-2} meV (Rigamonti and Caretta, 2009; Bransden and Joachain, 2000), which is 23.8 GHz or a wavelength of 1.26 cm. Equations (2.4.11) and (2.4.12) show how a non-uniform electric field may be used to separate a beam of ammonia molecules, according to whether the NH_3 is in the upper or the lower energy level of an inversion doublet. The needed apparatus is sketched in Fig. 4.3. The resonant cavity has provision for the input of microwaves, for the extraction of microwaves and for the monitoring of the power level in the resonant cavity. The apparatus in Fig. 4.3 is set to send the ammonia molecules in the upper energy level into the resonant cavity. Thus, we have a population inversion in the molecular beam that enters the resonant cavity. Within this beam, there are more ammonia molecules in the more energetic state than in the less energetic state of the inversion doublet. This permits stimulated emission from the upper energy level to the lower energy level. The frequency of the input microwaves is varied until it is in resonance with the doublet's transition frequency. The measured power level in the resonant cavity indicates when resonance and stimulated emission are achieved. The final result is coherent emission of the microwave radiation, hence, a maser.

The resonant cavity length is selected so that the ammonia molecule emits once and exits (Sakurai and Napolitano, 2011, p. 344).

Ammonia maser schematic

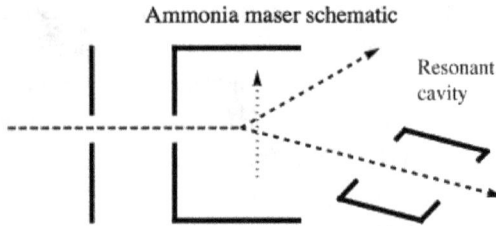

Figure 4.3. A schematic of an ammonia maser. The beam of NH_3 molecules enters from the left and is separated by a non-uniform electric field represented by the dotted arrow. The NH_3 molecules in the upper energy state then enter a resonant cavity with a time-varying electric field, which induces emission and the return of the NH_3 molecules to their lower energy state.

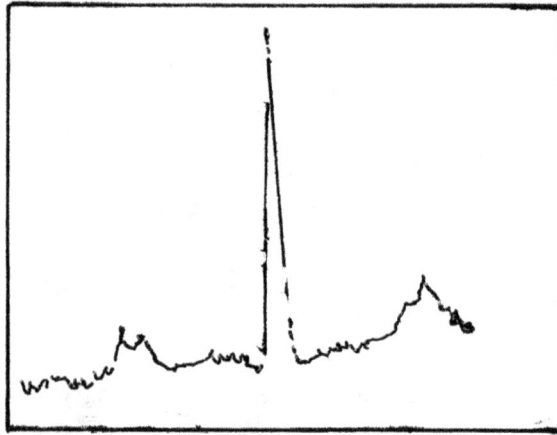

Figure 4.4. A tracing of the output of the ammonia maser (Gordon *et al.*, 1954, Fig. 2). The original figure is a white line on a black background. The frequency increases to the left. The secondary structures are about 50 kHz from the center peak at 23.8 GHz.

This corresponds to the π-pulse mentioned in Sec. 3.3, where the occupation probabilities are worked out for the two-level system starting in the upper energy level. Figure 4.4 represents the ammonia maser's output versus frequency. Gordon *et al.* (1954) estimate a full width at half maximum of 6 to 8 kHz for the 23.8 GHz line.

Maser action is also observed in interstellar space and is ascribed to molecular masers. For example, radio astronomers have observed ammonia masers (Madden *et al.*, 1986; Walmsley, 1994). They look for narrow emission line widths and intensities much stronger than related emission lines.

One aspect of the ammonia maser bears further consideration here. A two-level system is used, but illuminated two-level systems cannot maintain a population inversion. This is because an individual two-level system goes from the upper energy level to the lower energy level by the emission of a photon. This photon may be absorbed by a two-level system in the lower energy level, which means the photon is not emitted by the maser. The rate equation argument of Sec. 1.5 shows that the long-time population of the upper energy level is at best equal to that of the lower energy level. The use of a beam of ammonia molecules sidesteps this problem. Fresh ammonia molecules in the upper energy level are continually introduced into the resonant cavity, which also has an outlet to exhaust the ammonia molecules. Maser development turned to 3- and 4-level systems that allow population inversion. Such masers are discussed in Wilson and Hawkes (1989, Sec. 5,4) and Hecht (2005, Chap. 2). Maser research continues as witnessed by a micromaser involving quantum dots and single electron tunneling events (Liu *et al.*, 2015). However, we now shift our focus and next consider magnetic resonance.

4.3. Magnetic Resonance

Two-level systems are frequently used for magnetic resonance measurements. The occurrence of a resonance is used to provide the energy difference between the two energy levels associated with the spin of an electron or nucleon. This is useful in determining the structure and/or the constituents of a sample. In addition, values of the magnetic moment of the state or of the magnetic field may be obtained if the energy separation is known from other measurements. When electron spins are involved, the technique is called electron spin resonance or electron paramagnetic resonance. The general idea is to measure the amount of power absorbed from the time-varying magnetic field. This field is usually at microwave or radio frequencies. Ideally, the absorbed power goes into inducing transitions between the spin states.

Magnetic resonance is done quite frequently with liquids and solid samples by methods that differ from the molecular beam approach described below. Felix Bloch, Edward M. Purcell and Eugenii K. Zaroisky independently carried out the first experiments on magnetic resonance in condensed matter. Bertolotti (2005, pp. 160–168) provides a detailed account of the early work. Magnetic resonance is now a highly-developed tool for the determination of the structure of condensed matter. This may include both the identity of the atoms in a sample and their location with respect to each other. Comprehensive descriptions of magnetic resonance measurements are contained in Slichter (2010) and Serdyuk *et al.* (2007). Shorter summaries are provided in Bransden and Joachain (2000, Chap. 12) and Melissinos and Napolitano (2003, Chap. 7).

We start with an account of how the magnetic resonance appears and follow with a discussion of the early experiments of Rabi and colleagues. Generally, there is a constant magnetic field B_0 in one direction, say the z-direction, and a perpendicular magnetic field of magnitude B_1. The latter magnetic field is oscillatory. This resembles the case studied in Sec. 3.6 for a spin-$\frac{1}{2}$ system. We assumed an electron there. Here we consider a nucleon. Thus, we need an overall minus sign in the Hamiltonian, which is given by Eq. (3.6.2). We work in the Schrödinger Picture and we follow the Sec. 3.6 derivations for a rotating magnetic field. We take the $|-\rangle$ to be the lower energy state and $|+\rangle$ to be the higher energy state. The difference in the z-component of angular momentum between these two states is ± 1 and this is one of the selection rules for magnetic dipole transitions as shown in Cohen-Tannoudji *et al.* (1977, Complement A_{XIII}). The Hamiltonian of Eq. (3.6.2) results in the occupation probabilities of Eqs. (3.6.51) and (3.6.55). The latter applies to the excited state when the system is started in the upper energy state $|+\rangle$ at time $t = 0$. (When the system is initialized in the $|-\rangle$ state, we interchange the two occupation probabilities.)

We set

$$\Omega_0 = \omega_+ - \omega_-, \tag{4.3.1}$$

which is the difference in angular frequencies between the spin-up state $|+\rangle$ and the spin-down state $|-\rangle$. Each term in Eq. (4.3.1) is proportional to B_0 through the Zeeman effect (Konishi and Paffuti, 2009, Sec. 15.8). We note Ω_0 is also defined through Eq. (3.6.3). If the gyromagnetic ratio is negative, as for electrons or neutrons, then we use $-\Omega_0$. Equation (3.6.4) defines $\tilde{\Omega}$, which is proportional to the magnitude of the time-varying field B_1. In addition, when the angular frequency of the time-varying magnetic field is ω, we have

$$\delta\omega = \omega - \Omega_0, \tag{4.3.2}$$

with $\delta\omega = 0$ for resonance.

Magnetic resonance is concerned with the absorption of a photon of angular frequency ω and it helps to have a maximal population in the lower energy state, as this increases the amount of absorption. The maximum occupation probability of $|-\rangle$, $p_-(t)$, from Eq. (3.6.51) is

$$p_{-\,\text{max}} = \frac{\tilde{\Omega}^2}{\tilde{\Omega}^2 + (\delta\omega)^2}. \tag{4.3.3}$$

We next put

$$\Omega_0 = \gamma\omega, \tag{4.3.4}$$

because we consider ω to be fixed and the magnitude of the constant magnetic field B_0 (and hence Ω_0) to be the experimental variable. We rewrite Eq. (4.3.3) in terms of γ

$$p_{-\,\text{max}} = \frac{1}{1 + \left[\left(\frac{\omega}{\tilde{\Omega}}\right)(1-\gamma)\right]^2}, \tag{4.3.5}$$

and treat $\omega/\tilde{\Omega}$ as a parameter. Figure 4.5 shows how $p_{-\,\text{max}}$ varies with γ. The curves are Lorentzians and show a resonance when $\gamma = 1$, which is when $\Omega_0 = \omega$. Here the angular frequency of the time-varying magnetic field equals the separation in angular frequency between the states $|+\rangle$ and $|-\rangle$, hence, their energy separation is $\hbar\Omega_0$. Figure 4.5 shows the resonance is narrower for the larger value of $\omega/\tilde{\Omega}$. This corresponds to a higher ω or a smaller B_1.

Rabi *et al.* (1938a; 1938b) first demonstrated magnetic resonance after theoretical work by Rabi (1937). Rabi *et al.* (1939) provide a more complete description of the experiments. They used a beam of

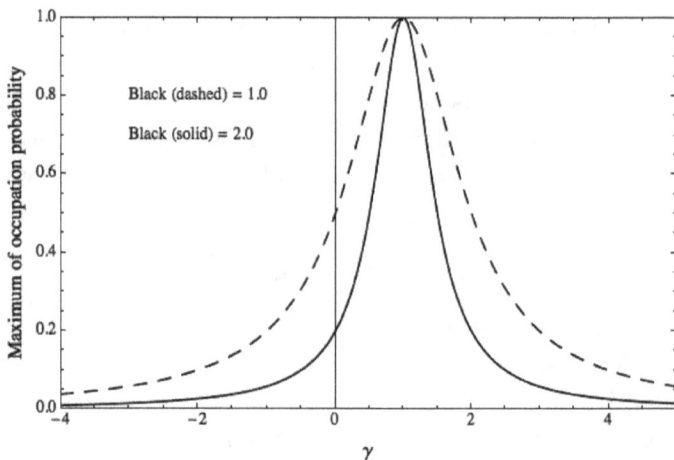

Figure 4.5. Maximum of occupation probability of the lower energy level versus γ defined in Eq. (4.3.4). Results are given for two values of $\omega/\tilde{\Omega}$, dashed = 1.0, solid = 2.0.

either LiCl or NaF molecules in a state such that the effects of the electronic spin could be ignored. The experiment probed the nuclear magnetic moment of Li or F. A schematic of their apparatus appears in Fig. 4.6 (Rabi *et al.*, 1939, Fig. 2). The beam exits the source, the oven O, and passes through a slit and then between the pole pieces of magnet A with a variation in the magnetic field in the z-direction. This splits the beam in an analogous manner to the separation of ammonia molecules in the inhomogeneous electric field described in Sec. 4.2. Let us call the beam with the spin-up particles B^+ and the beam with spin-down particles B^-. The beams then pass through a slit S and enter magnet C. This has a uniform magnetic field B_0 in the z-direction and a perpendicular magnetic field due to R that varies with time, our B_1 at angular frequency ω. It is here that the molecules in the beams B^+ and B^- make transitions between the $|+\rangle$ and $|-\rangle$ states. The beams then enter the third magnetic field in magnet B, which is similar to the first, but oppositely oriented. Upon exiting this third magnetic field, some of the molecules are detected after a slit at D. We show their results for fluorine with a nuclear spin of $\frac{1}{2}$ in Fig. 4.7.

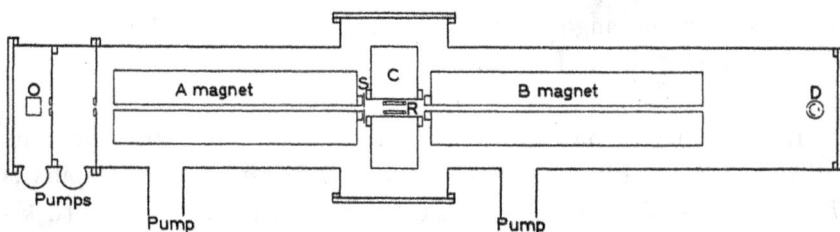

Figure 4.6. The apparatus of Rabi *et al.* (1939, Fig. 2). The molecular beams go from the left to the right. The letters are explained above.

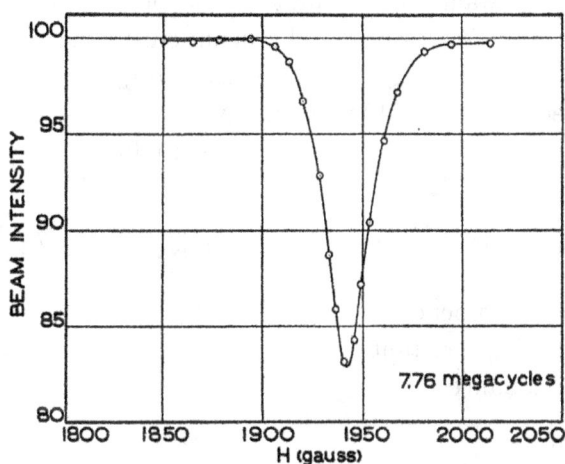

Figure 4.7. The measured results for F^{19} in NaF from Rabi *et al.* (1939). This is their Fig. 5. The constant magnetic field is varied and we divide by 10,000 to find the field in teslas.

Now if B_0 is zero and the time-varying magnetic field B_1 is off, then the two beams B^+ and B^- rejoin and the detector registers a strong signal. However, if $B_0 \neq 0$ and the time-varying magnetic field is on with ω close to the Ω_0 of Eq. (4.3.1), then some of the molecules in both beams will change their state. That is, they will flip their spin. Such molecules are then deflected by the third magnetic field and they do not rejoin the beam that passes through the slit D before the detector. Thus, the detected signal drops and, in fact, is a minimum at the resonance with $\Omega_0 = \omega$. Figure 4.7 is an example of the results of Rabi *et al.* (1939) for the fluorine nucleus. Here B_0

is varied and the angular frequency ω of B_1 is fixed. This curve is a mirror image of the Lorentzian of Fig. 4.5 or it may be viewed as $1 - p_{-\max}$.

It is useful to put some numbers into this account with the help of App. 1. First, we assume the temperature of the spins is $T = 300\,\mathrm{K}$, so $k_B T = 0.0258582\,\mathrm{eV}$ is the thermal energy with k_B equal to Boltzmann's constant. We consider the case shown in Fig. 4.7 (Rabi *et al.*, 1939, Fig. 5). The frequency $\nu = \omega/2\pi$ is 7.76 MHz and resonance for ^{19}F occurred at approximately 1940 gauss or 0.194 tesla. We want to look first at the populations of the spin-up and spin-down states before the beam enters the first magnet of Fig. 4.6. The lower energy state $|-\rangle$ has energy $E_- = -\hbar\Omega_0/2$ and the higher energy state $|+\rangle$ has energy $E_+ = +\hbar\Omega_0/2$. In thermal equilibrium, we expect the populations to follow the Boltzmann factors,

$$N = N_- + N_+ = A(Ne^{+\hbar\Omega_0/2k_B T} + Ne^{-\hbar\Omega_0/2k_B T}), \qquad (4.3.6)$$

where N is the number of molecules in a unit volume at a time t and A is a normalization constant that ensures the equality in Eq. (4.3.6).

Now at resonance

$$\hbar\Omega_0 = h\nu$$
$$= (6.626 \times 10^{-34}\ \text{joule-s})(7.76 \times 10^6/\text{s}) = 5.142 \times 10^{-27}\ \text{J}$$
$$= 3.21 \times 10^{-8}\ \text{eV}.$$

We check this number by noting that Fig. 4.7 indicates resonance occurs for $B_0 = 0.194$ teslas. Then

$$\hbar\Omega_0 = g\mu_N B_0$$
$$= 5.243(5.05078 \times 10^{-27}\ \text{J/T})(0.194\ \text{T}) = 5.137 \times 10^{-27}\ \text{J}.$$

The result is consistent and we have used the gyromagnetic ratio for ^{19}F (Rabi *et al.*, 1939, Table II) along with the nuclear magneton.

At 300 K, we can safely approximate the exponentials in Eq. (4.3.6) by their first two terms. And, in fact, this approximation should hold for a wide range of temperatures. Then, $A = 1/2$ and the population difference between the two states is

$$N_{diff} = N_- - N_+ = \frac{N}{2}(e^{h\nu/2k_BT} - e^{-h\nu/2k_BT}) = \frac{Nh\nu}{2k_BT}. \quad (4.3.7)$$

With the above numbers,

$$N_{diff}/N = \hbar\Omega_0/2k_BT = 0.62 \times 10^{-6}, \quad (4.3.8)$$

we see that the two states are nearly equally populated. At first, this seems at odds with the strong resonance feature in Fig. 4.7. So we consider the spin-up particles. Some leave beam B^+ when their spin flips, but these then have trajectories that keep them from joining beam B^-. Similarly, spin-down particles that flip their spin do not join beam B^+. Thus, the signal strength at the detector drops for both types of spin flip.

How many particles undergo a spin flip depends on two factors. The first is how near Ω_0 is to ω and the second is how many particles pass through the second magnet. If the separation $|\Omega_0 - \omega|$ is significant, then very few particles change their spin state. The situation changes dramatically as the separation is decreased and resonance is approached. The beams must pass through this second magnet slowly, with the passage time long enough to allow the occupation probabilities to change significantly. Ideally, we know ω, Ω_0 and $\tilde{\Omega}$, so we may use Eq. (3.6.51) and its counterpart Eq. (3.6.55) for the particles starting in $|+\rangle$, to provide an estimate of the time needed for maximum spin flipping to occur. B_0 is varied and the use of molecular beams allows a fresh supply of molecules for each B_0 value. Eventually, a plot such as Fig. 4.7 emerges.

In condensed matter, the sample replaces the molecular beam. Now the density of spins is increased tremendously in the sample and we need to worry about further interactions. For example, the spins may change state through interactions with other spins and through interactions with phonons, that is, with vibrations of the lattice for a solid. The Time-Dependent Schrödinger Equation of Sec. 3.6 needs to be generalized to handle these "relaxation" phenomena. This is often done with a version of the Bloch Equations that we develop in Chap. 5 after the density matrix is introduced. For now, we note that our sample of condensed matter needs to relax between

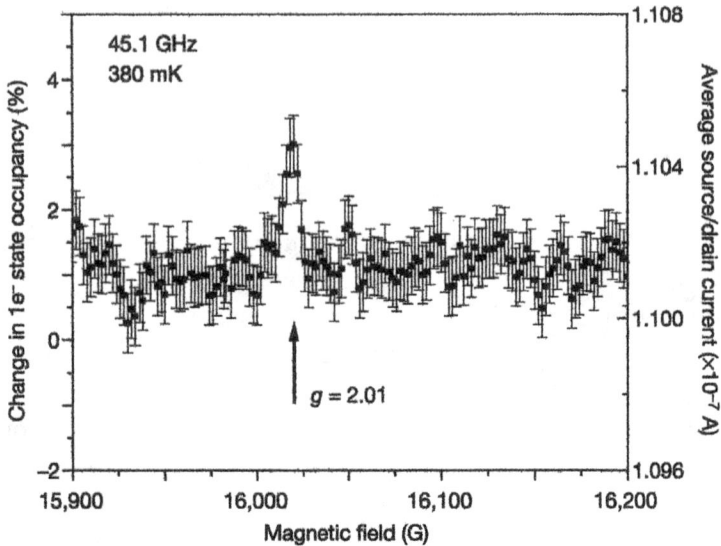

Figure 4.8. The left-hand axis has the change in the single electron state occupancy and a resonance peak is observed. The right-hand axis shows the source-drain current. The magnetic field needs to be divided by 10,000 to get teslas. Figure 4(a) of Xiao *et al.* (2004) is reprinted by permission from Macmillan Publishers Ltd: Nature, copyright 2004.

measurements in order for us to build up a curve analogous to that of Fig. 4.7.

An example of work in condensed matter is provided by Xiao and colleagues. They studied a single electron paramagnetic spin center due to a defect in the gate oxide of a silicon field-effect transistor. This defect location is suggested because the observed gyromagnetic ratio is greater than 2 (Xiao *et al.*, 2004). In addition, the defect needs to be adjacent to the conducting channel in the silicon in order for the defect to influence the source-drain current. The work is an extension of the random telegraph noise investigations discussed in Ramble 3 of Chap. 2. Figure 4.4(a) of Xiao *et al.* (2004) is reproduced here as Fig. 4.8. The resonance occurs at 1.6025 tesla when the microwave frequency is 45.1 GHz. This corresponds to the transition energy of 2.99×10^{-23} J $= 1.86 \times 10^{-4}$ eV. This explains why the measurements were done at 0.380 K, where $k_B T = 3.28 \times 10^{-5}$ eV. Xiao and colleagues also present limited data that show how the magnetic field for resonance varies with a change in the microwave frequency.

Section 4.4 continues with two-level defects found in diamond.

4.4. The NV$^-$ Defect Center in Diamond

Defects in diamond provide a variety of few-level systems such as the nitrogen-vacancy (NV) center discussed here. Since diamond is transparent to visible light, the defects can be probed optically. Kurtsiefer *et al.* (2000) have studied NV centers with the aim of developing a controllable single photon source, for which quantum dots are also considered (Matthiesen *et al.*, 2012). Here we explore the work of Fuchs *et al.* (2009) and their investigation of spin flipping at room temperature in a single NV center in diamond. Their experimental work is complemented by numerical calculations that go beyond the Rotating Wave Approximation (RWA). Our numerical solutions of Eqs. (3.2.2) and (3.2.3) also avoid the RWA and do capture the qualitative features found by Fuchs and colleagues. Their spin flipping or Rabi oscillations involve microwave frequencies, while optical excitation and non-radiative relaxation are used to put the NV center in the appropriate energy state for the Rabi oscillations. Further optical excitation is used to deduce the occupation of the spin-flipping states and we have an example of optically-detected magnetic resonance. This NV center involves more than two levels and we need to explain how a two-level result occurs and why room temperature works. This helps us understand why the NV center has been shown to be capable of sensing individual spins and fields. These two topics are examples of quantum sensing and are touched on briefly below.

The band gap of diamond is 5.5 eV as indicated in Fig. 4.9. The NV center is composed of a nitrogen atom adjacent to a lattice vacancy. The 3 carbon atoms around the vacancy each contribute an electron and the nitrogen atom provides 2 electrons. A sixth electron is captured from the bulk of the diamond. Thus, this is a NV$^-$ defect with a charge of -1. Doherty *et al.* (2011; 2012; 2013) discuss how the defect's electronic structure may be viewed in terms of molecular orbitals. This viewpoint shows two of the electrons occupying delocalized molecular orbitals within the diamond valence

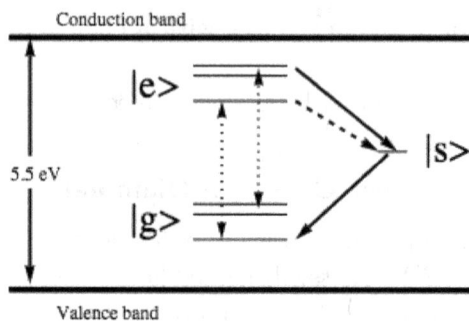

Figure 4.9. The NV$^-$ energy levels in diamond. This is based on Schirhagl *et al.* (2014, Fig. 4(a)). Optical transitions are indicated by dotted arrows. The dashed and solid arrows are non-radiative transitions. The former is the less likely transition. The state labels and the zero-field splitting are explained in the text. The energy level spacings are not to scale.

band (Doherty *et al.*, 2012). The remaining 4 electrons lead to levels within the diamond band gap, which are discussed in terms of spin-1 states. The C_{3v} symmetry of the defect helps in describing the energy states. The subscript 3 indicates rotational symmetry. We have three carbon atoms in a plane, so we have rotations of 0, $2\pi/3$, $4\pi/3$ and 2π. The vacancy and the nitrogen atom lie on a line perpendicular to the plane. The subscript v is for reflection in a plane. Here, it is a plane with the NV axis and one of the 3 carbon atoms. These symmetry operations of the group C_{3v} are well illustrated in Doherty *et al.* (2013, App.).

The NV$^-$ energy levels within the 5.5 eV energy gap of diamond are shown in Fig. 4.9 and their ordering is attributed to spin-spin interactions within the spin-1 system. The C_{3v} symmetry provides the wave functions, which are derived in Doherty *et al.* (2011). Their notation is explained in Haken and Wolf (1995, Chap. 6) and Doherty *et al.* (2013, App.). The ground state $|g\rangle$ is a spin triplet denoted as $^3A_2$. The zero-field splitting has the $m_S = 0$ state 2.87 GHz below the degenerate $m_S = \pm 1$ states. When a static magnetic field is applied parallel to the NV axis, the Zeeman effect (Konishi and Paffuti, 2009, Sec. 15.8) splits the $^3A_2$ states in the ways discussed below. The excited state $|e\rangle$ is a spin triplet called ^3E, while the state $|s\rangle$ is actually the two singlets ^1E and $^1A_1$. Further details and explanations of the NV energy levels are available in Doherty *et al.* (2011; 2012; 2013) and Schirhagl *et al.* (2014).

We next discuss why the NV$^-$ defect is useful at room temperature and we touch on the energy level ordering seen in Fig. 4.9. The latter eventually leads to the two-level system. The thermal energy at $T = 300\,\text{K}$ is $0.0258582\,\text{eV}$. We compare this with the zero-field splitting in the NV$^-$ ground state,

$$h(2.87 \times 10^9) = 1.19 \times 10^{-5} eV. \tag{4.4.1}$$

Thus, we expect the occupation probabilities of the $m_S = \pm 1$ and $m_S = 0$ states to be very similar to the magnetic resonance case of Sec. 4.3. However, we need to investigate how the spin is flipped thermally. Childress *et al.* (2014) point out that in diamond there is a small spin-orbit interaction and the NV spin is weakly-coupled to lattice phonons. The inset to Kolkowitz *et al.* (2015, Fig. 1D) shows relaxation times of milliseconds for a NV center in bulk diamond. We need an atom that is not ^{12}C for the spin flip, hence, we consider N and ^{13}C. Another N atom is generally hundreds of lattice spacings away, so the nuclear spins of naturally-occurring ^{13}C atoms are the main contributors to thermal spin flips. But this carbon isotope is only present at the 1.11% level and we have long relaxation times compared to the nanosecond periods of the exciting radiation. This allows the gigahertz microwave radiation in Fuchs *et al.* (2009) to control the occupation probabilities of the ground state spin states.

The explanation of the energy level ordering is equally involved. Schirhagl *et al.* (2014, Sec. 4.1) present an approximate Hamiltonian.

$$\hat{H} = \hbar D \left(\hat{S}_z^2 - \frac{2}{3} \right) + \hbar \tilde{\gamma} \bar{B} \cdot \hat{\bar{S}}. \tag{4.4.2}$$

The first term is based on the dipolar interactions between two spin-$\frac{1}{2}$ electrons and the interaction between the magnetic field and the total spin of the two electrons, which is 1 for the two triplet states of Fig. 4.9. This term may be derived from dipole-dipole interactions in coordinate space, along with the introduction of parameters, such as

D, that represent averages over the electron wave function. A good dose of spin algebra is required and Rieger (2007, Sec. 6.2) carries out a detailed derivation. The second term of Eq. (4.4.2) represents the Zeeman effect. We assume the static magnetic field is along the z-axis, which we take to be along the NV axis. The parameters are the $D = 2.87\,\mathrm{GHz}$ used in Eq. (4.4.1) and $\tilde{\gamma} = 28\,\mathrm{GHz/T}$, which, as expected, is the gyromagnetic ratio times the Bohr magneton in frequency units. Expressions for electric fields and for strain fields are often added to Eq. (4.4.2) and the appropriate terms are developed by, for example, Maze *et al.* (2011).

We use these values of D and $\tilde{\gamma}$ to check the energy level separations of Fuchs *et al.* (2009). The zero-field splitting is 2.87 GHz as they state, which we have depicted in Fig. 4.9. They use a static magnetic field of 0.085 tesla to induce the Zeeman splitting of the $^{3}\mathrm{A}_2$ states of $|g\rangle$. The $m_S = +1$ and $m_S = -1$ states are now 4.76 GHz apart and the $m_S = -1$ state is 0.49 GHz above the $m_S = 0$ ground state. Fuchs and colleagues use a microwave-frequency driving field to induce transitions between the $m_S = -1$ and the $m_S = 0$ states of $^{3}\mathrm{A}_2$. We see these two levels are well separated from the $m_S = +1$ state, hence, the $m_S = -1$ and the $m_S = 0$ states are viewed as a pseudospin $= \frac{1}{2}$ two-level system. And we are almost ready to return to the experimental data on the Rabi oscillations of the NV^- defect center.

We first explain how the $m_S = 0$ ground state is populated. The isolated NV center is optically-excited to the higher energy spin triplet in Fig. 4.9 by a pulse from a laser operating at 532 nm, which is green light at 2.33 eV. The NV center may emit an optical photon, so the excited electron transfers from the excited $m_S = 0$ state to the $m_S = 0$ state of the lower energy spin triplet. Similarly, optical transitions are possible between the excited state $m_S = \pm 1$ level and the ground state $m_S = \pm 1$ level. However, the excited $m_S = \pm 1$ states also relax strongly by non-radiative transitions to the singlet states at the far right of Fig. 4.9 and then on to the $m_S = 0$ state. This method of populating the $m_S = 0$ ground state is an example of optical spin polarization. The excited state $m_S = 0$ level is less likely to decay to the singlet states. With the laser turned off, a resonant microwave pulse with a selected pulse width then induces Rabi oscillations between the $m_S = 0$ and $m_S = -1$ states of the ground state spin triplet. Finally, the laser is turned back on and the photoluminescence from the excited triplet states is measured. The photoluminescence intensity, I_{PL}, is linearly dependent on the population p of the $m_S = 0$ ground state before the laser is turned back on. Fuchs *et al.* (2009) carefully calibrated their system and determined when the population p is equal to 1 ($m_S = 0$ state is occupied) and when $p = 0$ ($m_S = -1$ is occupied). This allowed them to map out the changes in the population as a function of the microwave pulse width.

This situation corresponds to Sec. 3.4, where the equations are reduced to those of Sec. 3.2. The resulting population of the ground state is the $p_1(t)$ given in Eq. (3.2.36). We needed the RWA to get this result. However, Fuchs and colleagues numerically solve equations like Eq. (3.4.8), and this takes them beyond the RWA. Figure 4.10 shows how their calculations reproduce their experimental data with Gaussian-shaped microwave pulses. Here the amplitude of the microwave pulse is expressed as a frequency through the analog of Eqs. (3.2.4) and (3.2.7), which are identical at resonance. As the amplitude approaches $0.49\,\text{GHz} = 490\,\text{MHz}$, which is the $m_S = 0$ to $m_S = -1$ transition frequency with the applied static magnetic field, the data acquire multiple oscillation frequencies. We see these in the calculations presented in Sec. 3.3.1. Shore (1990, Vol. 1, Sec. 3.8) discusses similar high-frequency behavior and Cohen-Tannoudji *et al.* (1977) provide an exposition in their Complement B to Chap. 13. In addition, Silverman and Pipkin (1972) find added oscillations in their calculated results for an atom interacting with a time-dependent electric field. Their Fig. 5 is particularly illustrative.

These results of Fuchs *et al.* (2009) show they are able to accurately simulate their single spin NV center at room temperature. This lets them evaluate how quickly they are able to controllably flip a spin and emit a photon. This ability is of interest for on-chip communication in integrated circuits, quantum information processing or quantum computing. These areas continue to be the subjects of intense research and development.

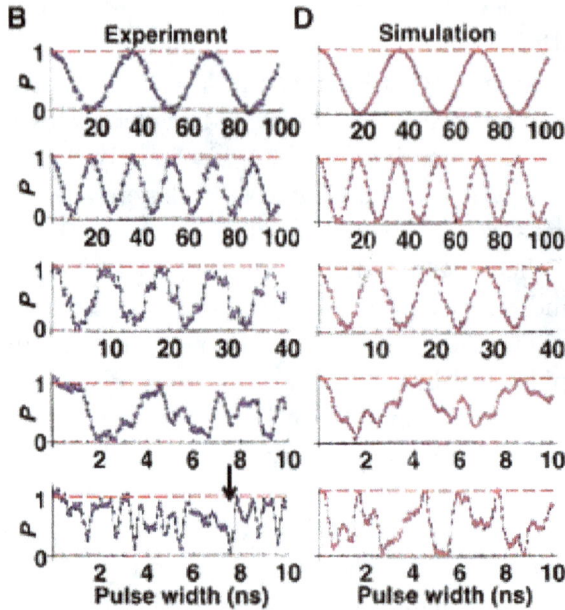

Figure 4.10. Rabi oscillations: experimental data (B) versus simulations (D) (Fuchs *et al.*, 2009, Figs. 2B and 2D). The amplitudes of the microwave pulse from the top trace downward are: 29, 57, 109, 223 and 440 MHz, respectively. Please note the changes in the time scale. Figure from *Science* **326** (2009). Reprinted with permission from AAAS.

Before we leave this topic, we show how the numerical solutions of Eqs. (3.2.2) and (3.2.3) behave. This fuels a comparison with the data in Fig. 4.10 and was the original reason to include the Rabi oscillations of the NV^- defect center. However, as the above shows, this contact with NV^- was only the "tip of an iceberg". We do not make the RWA, but we do assume we are at resonance, hence, the angular frequency is 2π (0.49 GHz). We numerically solve the resulting equations with the routine NDSolve of Mathematica (Version 9, 2012). The intent is to demonstrate that these numerical solutions capture the qualitative behavior of the more complete calculations of Fuchs *et al.* (2009) that use a pulse. The intensity of the oscillating magnetic field of the microwave pulse is adjusted to match the 29 MHz results of Fuchs and colleagues. The results of our calculations appear in Fig. 4.11 and closely resemble the 29 MHz

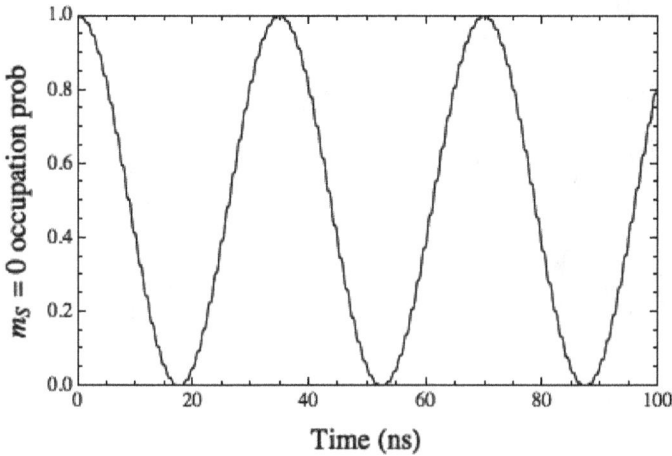

Figure 4.11. $p(t)$, the occupation probability of the $m_S = 0$ ground state, versus the length of the microwave pulse for the base pulse intensity.

data. The microwave pulse intensity is then scaled by 4 and 16 for Figs. 4.12 and 4.13, respectively. The trends of Fig. 4.10 are clearly seen, that is, the change in the periodicity and the development of higher frequency components are tracked with the increase in the microwave pulse intensity.

We close this section with comments on how NV$^-$ defect centers are being applied to "sense" or measure spins and fields in condensed matter. The idea is to use a single defect center to probe on a micro or nano scale. Childress *et al.* (2014) provide a brief introduction to quantum sensing, while Schirhagl *et al.* (2014) describe several applications in detail. The latter generalizes the Hamiltonian of Eq. (4.4.2) to treat electric fields interacting with spins. Maze *et al.* (2011) extend the Hamiltonian to also include strain. The basic idea is to explore the response of the ground state triplet of the NV$^-$ defect center to magnetic and electric fields or to strain. The changes in the level spacings are determined through optical and microwave probes. The literature on quantum sensing is growing rapidly as more uses for NV$^-$ defect centers are found.

We next turn to Sec. 4.5 and consider the Pulse Area Theorem that is used in Sec. 4.1.

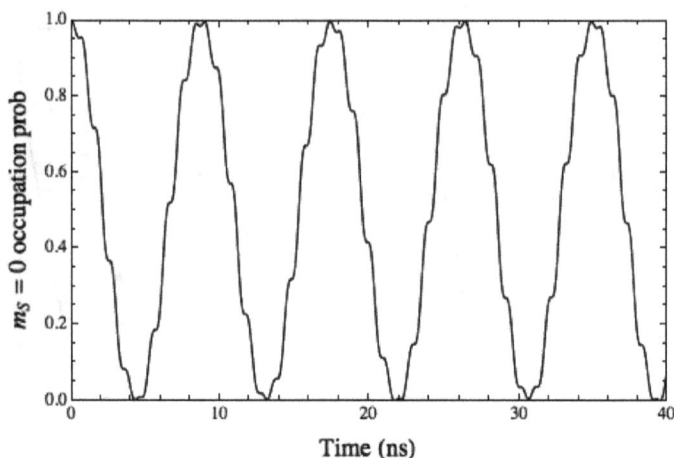

Figure 4.12. $p(t)$ versus the length of the microwave pulse for 4 times the base pulse intensity.

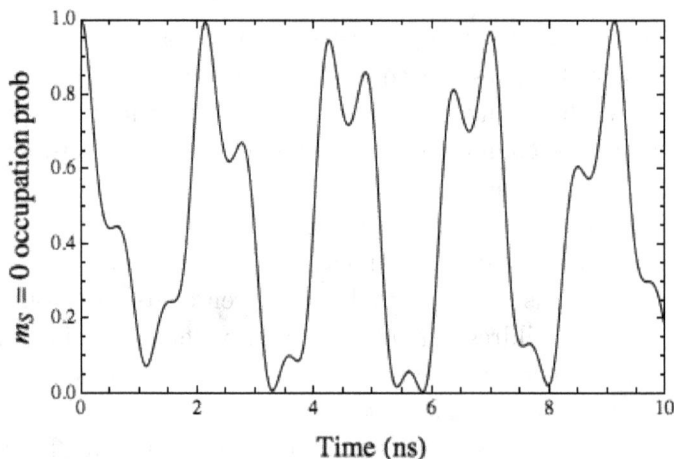

Figure 4.13. $p(t)$ versus the length of the microwave pulse for 16 times the base pulse intensity.

4.5. Ramble 4: The Pulse Area Theorem

Pulse areas appear in Sec. 4.1, so it is worth outlining the Pulse Area Theorem. The result is found in Berman and Malinovsky (2011, p. 29) and Tannor (2007, pp. 481–482). We return to the coupled equations for the amplitudes in the Rotating Wave Approximation (RWA), Eqs. (3.2.5) and (3.2.6), which are,

$$\frac{dc_1(t)}{dt} = (i\tilde{\Omega}/2)e^{i\delta\omega t}c_2(t), \qquad (4.5.1)$$

$$\frac{dc_2(t)}{dt} = (i\tilde{\Omega}/2)e^{-i\delta\omega t}c_1(t). \qquad (4.5.2)$$

Here

$$\tilde{\Omega} = \mu E/\hbar, \qquad (4.5.3)$$

and $\delta\omega$ defines how far the angular frequency of the time-dependent electric field is from the difference in the angular frequencies of energy levels 2 and 1. We assume resonance, so $\delta\omega = 0$. Now we allow the strength of the electric field E to be dependent on the time, so we have $\tilde{\Omega}(t)$. This lets us approximate a pulse, which we assume to go from time 0 to t. Equations (4.5.1) and (4.5.2) become

$$\frac{dc_1(t)}{dt} = i\frac{\tilde{\Omega}(t)}{2}c_2(t), \qquad (4.5.4)$$

$$\frac{dc_2(t)}{dt} = i\frac{\tilde{\Omega}(t)}{2}c_1(t). \qquad (4.5.5)$$

We note that when $\tilde{\Omega}(t)$ is a constant,

$$\begin{aligned} c_1(t) &= \cos(\tilde{\Omega}t/2) \\ c_2(t) &= i\sin(\tilde{\Omega}t/2) \end{aligned}, \qquad (4.5.6)$$

as is verified by direct substitution. The generalization of these solutions is

$$c_1(t) = \cos\left(\int_0^t \frac{\tilde{\Omega}(t')dt'}{2}\right), \qquad (4.5.7)$$

$$c_2(t) = i\sin\left(\int_0^t \frac{\tilde{\Omega}(t')dt'}{2}\right). \qquad (4.5.8)$$

The time derivative of the integral provides one non-zero term and we find these solutions satisfy Eqs. (4.5.4) and (4.5.5).

Equations (4.5.7) and (4.5.8) lead to occupation probabilities that depend on the integral of $\tilde{\Omega}(t)$ and not the details of $\tilde{\Omega}(t)$. This allows us to analyze data such as that of Gibbs (1973), which plots the integrated fluorescence versus the input-pulse area.

References

P. R. Berman and V. S. Malinovsky, *Principles of Laser Spectroscopy and Quantum Optics* (Princeton University Press, Princeton, 2011).

M. Bertolotti, *The History of the Laser* (Institute of Physics Publishing, Bristol, 2005).

B. H. Bransden and C. J. Joachain, *Quantum Mechanics*, 2nd ed. (Prentice Hall, Harlow, England, 2000).

L. Childress, R. Walsworth, and M. Lukin, "Atom-like crystal defects", *Physics Today* **67**, October, 38–43 (2014).

C. Cohen-Tannoudji, B. Diu, and F. Laloë, *Quantum Mechanics*, Vols. 1 and 2 (John Wiley and Sons, New York, 1977).

M. W. Doherty, N. B. Manson, P. Delaney, and L. C. L. Hollenberg, "The negatively charged nitrogen-vacancy centre in diamond: The electronic solution", *New Journal of Physics* **13**, 025019 (2011).

M. W. Doherty, F. Dolde, H. Fedder, F. Jelezko, J. Wrachtrup, N. B. Manson, and L. C. L. Hollenberg, "Theory of the ground-state spin of the NV^{-} center in diamond", *Phys. Rev. B* **85**, 205203 (2012).

M. W. Doherty, N. B. Manson, P. Delaney, F. Jelezko, J. Wrachtrup, and L. C. L. Hollenberg, "The nitrogen-vacancy colour centre in diamond", *Physics Reports* **528**, 1–45 (2013).

R. P. Feynman, R. B. Leighton, and M. Sands, *The Feynman Lectures on Physics: Quantum Mechanics* (Addison-Wesley, Reading, Massachusetts, 1965).

M. Fox, *Quantum Optics: An Introduction* (Oxford University Press, Oxford, 2006), reprinted 2007.

G. D. Fuchs, V. V. Dobrovitski, D. M. Toyli, F. J. Heremans, D. D. Awschalom, "Gigahertz dynamics of a strongly driven single quantum spin", *Science* **326**, 1520–1522 (2009).

H. M. Gibbs, "Spontaneous decay of coherently excited Rb", *Phys. Rev. Lett.* **29**, 459–462 (1972).

H. M. Gibbs, "Incoherent resonance fluorescence from a Rb atomic beam excited by a short coherent optical pulse", *Phys. Rev. A* **8**, 446–455 (1973).

S. Gleyzes, S. Kuhr, C. Guerlin, J. Bernu, S. Deléglise, U. B. Hoff, M. Brune, J.-M. Raimond, and S. Haroche, "Quantum jumps of light recording the birth and death of a photon in a cavity", *Nature* **446**, 297–300 (2007).

J. P. Gordon, H. J. Zeiger, and C. H. Townes, "Molecular microwave oscillator and new hyperfine structure in the microwave spectrum of NH$_3$", *Phys. Rev.* **95**, 282–284 (1954).

J. P. Gordon, H. J. Zeiger, and C. H. Townes, "The Maser — new type of microwave amplifier, frequency standard, and spectrometer", *Phys. Rev* **99**, 1264–1274 (1955).

A. Goswami, *Quantum Mechanics*, 2nd ed. (Waveland Press, Long Grove, Illinois, 2003).

H. Haken and H. C. Wolf, *Molecular Physics and Elements of Quantum Chemistry: Introduction to Experiments and Theory* (Springer-Verlag, Berlin, 1995).

J. Hecht, *Beam The Race to Make the Laser* (Oxford University Press, New York, 2005).

S. Kolkowitz, A. Safira, A. A. High, R. C. Devlin, S. Choi, Q. P. Unterreithmeier, D. Patterson, A. S. Zibrov, V. E. Manucharyan, H. Park, and M. D. Lukin, "Probing Johnson noise and ballistic transport in normal metals with a single-spin qubit", *Science* **347**, 1129–1132 (2015).

K. Konishi and G. Paffuti, *Quantum Mechanics: A New Introduction* (Oxford University Press, Oxford, 2009).

C. Kurtsiefer, S. Mayer, P. Zarda, and H. Weinfurter, "Stable solid-state source of single photons", *Phys. Rev. Lett.* **85**, 290–293 (2000).

Y.-Y. Liu, J. Stehlik, C. Eichler, M. J. Gullans, J. M. Taylor, and J. R. Petta, "Semiconductor double quantum dot micromaser", *Science* **347**, 285–287 (2015).

S. C. Madden, W. M. Irvine, H. E. Matthews, R. D. Brown, and P. D. Godfrey, "New interstellar masers in nonmetastable ammonia", *Astrophysical Journal* **300**, L79–L84 (1986).

F. G. Major, *The Quantum Beat: The Physical Principles of Atomic Clocks* (Springer-Verlag, New York, **1998**).

Mathematica = Wolfram Research, Inc., Mathematica, Version 9.0.0.0, Champaign, IL (2012).

C. Matthiesen, A. N. Vamivakas, and M. Atatüre, "Subnatural linewidth single photons from a quantum dot", *Phys. Rev. Lett.* **108**, 093602 (2012).

J. R. Maze, A. Gali, E. Togan, Y. Chu, A. Trifonov, E. Kaxiras, and M. D. Lukin, "Properties of nitrogen-vacancy centers in diamond: The group theoretic approach", *New Journal of Physics* **13**, 025025 (2011).

J. D. McGervey, *Introduction to Modern Physics*, 2nd ed. (Academic Press, Orlando, 1983).

A. C. Melissinos and J. Napolitano, *Experiments in Modern Physics*, 2nd ed. (Academic Press, San Diego, 2003).

J. G. Muga, R. S. Mayato, and I. L. Egusquiza, eds., *Time in Quantum Mechanics*, 2nd ed. (Springer, Berlin, 2008).

I. I. Rabi, "Space quantization in a gyrating magnetic field", *Phys. Rev.* **51**, 652–654 (1937).

I. I. Rabi, J. R. Zacharias, S. Millman, and P. Kusch, "A new method of measuring nuclear magnetic moment", *Phys. Rev.* **53**, 318 (1938a).

I. I. Rabi, S. Millman, P. Kusch, and J. R. Zacharias, "The magnetic moments of $_3Li^6$, $_3Li^7$ and $_9F^{19}$", *Phys. Rev.* **53**, 495 (1938b).

I. I. Rabi, S. Millman, P. Kusch, and J. R. Zacharias, "The molecular beam resonance method for measuring nuclear magnetic moments: The magnetic moments of $_3Li^6$, $_3Li^7$ and $_9F^{19}$", *Phys. Rev.* **55**, 526–535 (1939).

P. H. Rieger, *Electron Spin Resonance: Analysis and Interpretation* (RSC Publishing, Cambridge, 2007).

A. Rigamonti and P. Carretta, *Structure of Matter: An Introductory Course with Problems and Solutions*, 2nd ed. (Springer-Verlag Italia, Milan, 2009).

J. J. Sakurai and J. Napolitano, *Modern Quantum Mechanics*, 2nd ed. (Addison-Wesley, Boston, 2011).

R. Schirhagl, K. Chang, M. Loretz, and C. L. Degen, "Nitrogen-vacancy centers in diamond: Nanoscale sensors for physics and biology", *Annual Review of Physical Chemistry* **65**, 83–105 (2014).

I. N. Serdyuk, N. R. Zaccai, and J. Zaccai, *Methods in Molecular Biophysics: Structure, Dynamics, Function* (Cambridge University Press, Cambridge, 2007).

B. W. Shore, *The Theory of Coherent Atomic Excitation*, Vol. 1 Simple Atoms and Fields (John Wiley and Sons, New York, 1990).

M. P. Silverman and F. M. Pipkin, "Interaction of a decaying atom with a linearly polarized oscillating field", *J. Phys. B: Atom. Molec. Phys.* **5**, 1844–1860 (1972).

C. P. Slichter, *Principles of Magnetic Resonance*, 3rd ed. (Springer-Verlag, New York, 2010).

D. J. Tannor, *Introduction to Quantum Mechanics: A Time-Dependent Perspective* (University Science Books, Sausalito, California, 2007).

J. S. Townsend, *A Modern Approach to Quantum Mechanics* (University Science Books, Sausalito, CA, 2000).

C. M. Walmsley, "Ammonia in the interstellar medium", in *Molecules and Grains in Space, AIP Conference Proceedings 312*, Irène Nenner, ed., 463–475 (1994).

J. Wilson and J. F. B. Hawkes, *Optoelectronics: An Introduction*, 2nd ed. (Prentice Hall, New York, 1989).

M. Xiao, I. Martin, E. Yablonovitch, and H. W. Jiang, "Electrical detection of the spin resonance of a single electron in a silicon field-effect transistor", *Nature* **430**, 435–439 (2004).

Chapter 5

The Density Matrix and the Relaxation of Two-Level Systems

5.1. Why the Density Matrix?

So far, we have considered transitions between the two-level system's ground state and excited state due to the absorption of energy from an external field. The subsequent decay to the ground state has been ascribed to induced emission by the same external field. As noted in Sec. 3, Chap. 2, once the external field is turned off, the occupation probabilities are independent of time, because we have assumed our two-level system exists in isolation. There are two remedies. First, in this chapter, we consider interactions between the two-level system and its environment. These effects will be included through the use of John von Neumann's density matrix approach. This allows us to incorporate relaxation effects that cause transitions between the two levels and take the system into thermal equilibrium. Second, we need to include spontaneous emission. We do not treat this topic here. Instead, we refer the reader to Cohen-Tannoudji *et al.* (1992, Sec. III-C). However, we do incorporate an approximate model for spontaneous emission in Chap. 6.

In the previous chapters we solved for the coefficients of the basis states of the wave function. For example, we determined $c_1(t)$ and $c_2(t)$ of Eq. (3.1.7). The density matrix is built up from products of these coefficients. This approach leads us directly to the occupation probabilities, as we show in Sec. 5.2. We consider two interactions that lead to relaxation for a spin-$\frac{1}{2}$ system. We have a spin-spin interaction when the spin interacts with another spin. When our spin-$\frac{1}{2}$ system is in a condensed phase of matter, then we may have a spin-lattice interaction. Both types of interactions introduce relaxation mechanisms that we need to incorporate into the density matrix. And, finally, stray fields also provide relaxation mechanisms for our two-level system. The stated relaxation mechanisms have the aspect of random perturbations, or fluctuations, that affect our specific two-level system. We treat these interactions within the density matrix approach through the introduction of phenomenological time constants. References to more complete treatments are given as we proceed. The incorporation of relaxation changes the occupation probabilities of the two levels when the external field is on and when it is off.

Section 5.2 defines the density operator $\hat{\rho}$ and the associated density matrix ρ. We also show that if we have the density matrix, we are able to compute the values of observables, such as the energy. The Liouville–von Neumann Equation for the time-development of ρ is derived in Sec. 5.3 and the density matrix for a two-level case is worked out in Sec. 5.4. Relaxation is introduced in Sec. 5.5 and an example with relaxation is solved in Sec. 5.6, where several assumptions are made in order to obtain a closed-form solution. Combinations of density matrix elements lead to the Bloch Equations in Sec. 5.7. The Bloch Equations are solved in Sec. 5.8 for the precession of the spin-$\frac{1}{2}$ angular momentum vector. Then, Sec. 5.9 provides leads in several directions for the further study of density matrices. Finally, Ramble 5 appears in Sec. 5.10. It is devoted to the solution of the time-dependent equations for the elements of the density matrix when an electromagnetic field is present.

Density matrices are discussed in many texts on Quantum Mechanics such as Sakurai and Napolitano (2011) and Bransden and Joachain (2000). In addition, Statistical Mechanics texts often treat density matrices. The following material has elements of Fayer (2001, Chap. 14) and Snoke (2009, Chap. 9) and a strong dose of Blum (1996, Chaps. 2 and 8). (Beware of frequent typographical errors in this edition of Blum.) Further insight is found in Schatz and Ratner (2002, Chap. 11), Bittner (2010, Chap. 6), Rand (2010) and Stenholm (2005). Here, Chap. 6 contains a closed-form solution of the Liouville–von Neumann Equations when relaxation is present, although computer algebra is required.

5.2. The Density Operator and the Density Matrix

In Chaps. 2 and 3, we solve the Schrödinger Equation for the coefficients of our complete set of states $\{|\phi_i\rangle\}$. The ket we sought was defined in Eqs. (2.1.8) and (3.1.7). We rewrite the latter now as

$$|\psi(t)\rangle = a_1(t)|1\rangle + a_2(t)|2\rangle, \tag{5.2.1}$$

with

$$\langle i|j\rangle = \delta_{ij}, \tag{5.2.2}$$

and i and $j = 1$ or 2 for our two-level system. The $|a_i(t)|^2$ give us the occupation probabilities of states 1 and 2 as functions of time. The normalization

$$\langle \psi(t)|\psi(t)\rangle = 1, \tag{5.2.3}$$

leads us, with Eq. (5.2.2), to

$$|a_1(t)|^2 + |a_2(t)|^2 = 1. \tag{5.2.4}$$

In general, $a_1(t)$ and $a_2(t)$ are complex, so each has a real and an imaginary part. This leads to 4 time-dependent functions, but Eq. (5.2.4) is a constraint, so we have only 3 independent time-dependent functions.

The density operator $\hat{\rho}$ provides the three independent functions and is an alternate approach to the direct solution of the Schrödinger Equation. We define $\hat{\rho}$ as a projection operator

$$\hat{\rho}(t) = |\psi(t)\rangle\langle\psi(t)|, \tag{5.2.5}$$

and the density matrix

$$\rho_{ij} = \langle i|\hat{\rho}|j\rangle. \tag{5.2.6}$$

Then, with Eqs. (5.2.1) and (5.2.2), we find,

$$\rho_{11} = \langle 1|(a_1|1\rangle + a_2|2\rangle)(a_1^*\langle 1| + a_2^*\langle 2|)|1\rangle = a_1 a_1^*, \tag{5.2.7}$$

and similarly,

$$\rho_{12} = a_1 a_2^*, \qquad (5.2.8)$$

$$\rho_{21} = a_2 a_1^*, \qquad (5.2.9)$$

$$\rho_{22} = a_2 a_2^*. \qquad (5.2.10)$$

We write the density matrix ρ as

$$\rho(t) = \begin{bmatrix} a_1(t)a_1^*(t) & a_1(t)a_2^*(t) \\ a_2(t)a_1^*(t) & a_2(t)a_2^*(t) \end{bmatrix}. \qquad (5.2.11)$$

The diagonal matrix elements $a_1 a_1^*$ and $a_2 a_2^*$ of $\rho(t)$ give the occupation probabilities of the states 1 and 2, respectively. The density matrix of Eq. (5.2.11) is seen to be Hermitian. The trace Tr of a matrix is the sum of its diagonal matrix elements. Thus,

$$\text{Tr}(\rho(t)) = a_1 a_1^* + a_2 a_2^* = 1, \qquad (5.2.12)$$

by Eq. (5.2.4). The off-diagonal matrix elements are known as the coherences and they will help us see how our two-level system returns to equilibrium after an external field is turned off.

The expectation value of an observable may be calculated if the elements of the density matrix are known. To see this, we start with the operator for the observable, for example, the Hamiltionian $\hat{H}$ and the energy E. Then,

$$E = \langle \psi(t)|\hat{H}|\psi(t)\rangle = (a_1^*\langle 1| + a_2^*\langle 2|)\hat{H}(a_1|1\rangle + a_2|2\rangle), \qquad (5.2.13)$$

$$E = a_1^* a_1 \langle 1|\hat{H}|1\rangle + a_2^* a_1 \langle 2|\hat{H}|1\rangle + a_1^* a_2 \langle 1|\hat{H}|2\rangle + a_2^* a_2 \langle 2|\hat{H}|2\rangle, \qquad (5.2.14)$$

and

$$E = \sum_{i=1}^{2}\sum_{j=1}^{2} a_i^* a_j \langle i|\hat{H}|j\rangle. \qquad (5.2.15)$$

We recognize the density matrix elements here with

$$a_i^* a_j = a_j a_i^* = \rho_{ji} = \langle j|\hat{\rho}|i \rangle. \tag{5.2.16}$$

Thus, we rewrite Eqs. (5.2.13) to (5.2.15) as

$$E = \langle \psi(t)|\hat{H}|\psi(t) \rangle = \sum_{i=1}^{2} \sum_{j=1}^{2} \langle j|\hat{\rho}|i \rangle \langle i|\hat{H}|j \rangle = \sum_{j=1}^{2} \langle j|\hat{\rho}\hat{H}|j \rangle,$$

$$\tag{5.2.17}$$

where the last equality follows because the $\{|i\rangle\}$ form a complete set of states. Finally, we see

$$E = \langle \psi(t)|\hat{H}|\psi(t) \rangle = \sum_{j=1}^{2} \langle j|\hat{\rho}\hat{H}|j \rangle = \mathrm{Tr}(\hat{\rho}\hat{H}). \tag{5.2.18}$$

This important result shows that the expectation value of an observable is gotten by computing the trace of the matrix product of the density operator and the operator of the observable. This relation holds for any observable. Hence, if we have the density matrix, we have the observable! The next step is to find an equation that leads to the density matrix and this we do in Sec. 5.3.

5.3. An Equation for the Time-Dependent Density Matrix

We need differential equations for the matrix elements of the density operator $\hat{\rho}(t)$. We start with the definition of the density operator $\hat{\rho}(t)$ in Eq. (5.2.5) and take the time derivative

$$\frac{d\hat{\rho}(t)}{dt} = \frac{d}{dt}|\psi(t)\rangle\langle\psi(t)| = \frac{d}{dt}(|\psi(t)\rangle)\langle\psi(t)| + |\psi(t)\rangle\frac{d}{dt}\langle\psi(t)|. \tag{5.3.1}$$

We use the Time-Dependent Schrödinger Equation

$$i\hbar\frac{d|\psi(t)\rangle}{dt} = \hat{H}|\psi(t)\rangle, \tag{5.3.2}$$

and its adjoint to rewrite Eq. (5.3.1) as

$$\frac{d\hat{\rho}(t)}{dt} = \frac{1}{i\hbar}\hat{H}|\psi(t)\rangle\langle\psi(t)| - \frac{1}{i\hbar}|\psi(t)\rangle\langle\psi(t)|\hat{H}. \tag{5.3.3}$$

We recognize the density operator on the right-hand side, so

$$i\hbar\frac{d\hat{\rho}(t)}{dt} = \hat{H}(t)\hat{\rho}(t) - \hat{\rho}(t)\hat{H}(t), \qquad (5.3.4)$$

where we allow for a time-dependent Hamiltonian. The right-hand side of Eq. (5.3.4) is the commutator, hence,

$$i\hbar\frac{d\hat{\rho}(t)}{dt} = [\hat{H}(t), \hat{\rho}(t)], \qquad (5.3.5)$$

and this equation, which is known as the Liouville or Liouville–von Neumann Equation, provides us with the time dependence of the density operator. We now need to get usable equations.

We explicitly assume the $\{|i\rangle\}$ are independent of time. This allows us to multiply Eq. (5.3.4) from the left by $\langle i|$ and from the right by $|j\rangle$,

$$i\hbar\frac{d\langle i|\hat{\rho}(t)|j\rangle}{dt} = \langle i|\hat{H}(t)\hat{\rho}(t)|j\rangle - \langle i|\hat{\rho}(t)\hat{H}(t)|j\rangle. \qquad (5.3.6)$$

The insertion of a complete set of states between the operators leads to

$$i\hbar\frac{d\langle i|\hat{\rho}(t)|j\rangle}{dt} = \sum_{k=1}^{2}\langle i|\hat{H}(t)|k\rangle\langle k|\hat{\rho}(t)|j\rangle - \sum_{k=1}^{2}\langle i|\hat{\rho}(t)|k\rangle\langle k|\hat{H}(t)|j\rangle,$$
$$(5.3.7)$$

for our two-level system. We write this out in terms of matrix elements

$$i\hbar\frac{d\rho_{ij}}{dt} = H_{i1}\rho_{1j} + H_{i2}\rho_{2j} - \rho_{i1}H_{1j} - \rho_{i2}H_{2j}. \qquad (5.3.8)$$

An examination of Eq. (5.3.8) shows this is matrix multiplication for 2×2 matrices. Hence,

$$i\hbar\frac{d}{dt}\begin{bmatrix} \rho_{11} & \rho_{12} \\ \rho_{21} & \rho_{22} \end{bmatrix} = \begin{bmatrix} H_{11} & H_{12} \\ H_{21} & H_{22} \end{bmatrix}\begin{bmatrix} \rho_{11} & \rho_{12} \\ \rho_{21} & \rho_{22} \end{bmatrix}$$
$$- \begin{bmatrix} \rho_{11} & \rho_{12} \\ \rho_{21} & \rho_{22} \end{bmatrix}\begin{bmatrix} H_{11} & H_{12} \\ H_{21} & H_{22} \end{bmatrix}. \qquad (5.3.9)$$

This is equivalent to the coupled set of equations

$$i\hbar\frac{d\rho_{11}}{dt} = H_{11}\rho_{11} + H_{12}\rho_{21} - \rho_{11}H_{11} - \rho_{12}H_{21} = H_{12}\rho_{21} - H_{21}\rho_{12},$$
$$(5.3.10)$$

$$i\hbar\frac{d\rho_{12}}{dt} = (H_{11} - H_{22})\rho_{12} + H_{12}(\rho_{22} - \rho_{11}), \qquad (5.3.11)$$

$$i\hbar\frac{d\rho_{21}}{dt} = H_{21}(\rho_{11} - \rho_{22}) + (H_{22} - H_{11})\rho_{21}, \qquad (5.3.12)$$

$$i\hbar\frac{d\rho_{22}}{dt} = H_{21}\rho_{12} + H_{22}\rho_{22} - \rho_{21}H_{12} - \rho_{22}H_{22} = H_{21}\rho_{12} - H_{12}\rho_{21},$$
$$(5.3.13)$$

where we use the commutativity of the matrix elements.

We note that with

$$\rho_{11} + \rho_{22} = 1, \qquad (5.3.14)$$

we expect

$$i\hbar\frac{d\rho_{11}}{dt} = -i\hbar\frac{d\rho_{22}}{dt}, \qquad (5.3.15)$$

and Eqs. (5.3.10) and (5.3.13) agree with this. In addition, since Eqs. (5.2.7) to (5.2.10) establish that $\hat{\rho}$ is a Hermitian operator, we have

$$\rho_{12} = (\rho_{21})^*. \qquad (5.3.16)$$

We substitute this into Eq. (5.3.11), take the complex conjugate of both sides of the equation, multiply by -1, and find Eq. (5.3.12). This result does need $\hat{H}$ to be Hermitian. In summary,

$$i\hbar\frac{d\rho_{12}}{dt} = i\hbar\frac{d\rho_{21}^*}{dt}. \qquad (5.3.17)$$

These constraints help us solve Eqs. (5.3.10) to (5.3.13).

As we have seen in the earlier chapters, many Hamiltonians separate into

$$\hat{H}(t) = \hat{H}_0 + \hat{V}(t), \qquad (5.3.18)$$

with all the time dependence in $\hat{V}(t)$. To take advantage of this separation, we need to modify the above derivation for the matrix elements of the density operator. We return to Eq. (5.3.5) and we now use the basis states that are orthonormal eigenkets of $\hat{H}_0$. Hence,

$$\hat{H}_0|i\rangle = E_i|i\rangle, \tag{5.3.19}$$

and we have shifted our notation from H_i to E_i for the eigenvalues of $\hat{H}_0$. Further,

$$\langle i|\hat{H}_0 + \hat{V}(t)|j\rangle = E_j\delta_{ij} + \langle i|\hat{V}(t)|j\rangle. \tag{5.3.20}$$

Now we are ready to multiply Eq. (5.3.5) on the left by $\langle i|$ and on the right by $|j\rangle$. We follow this with the insertion of complete sets of states. Equation (5.3.7) becomes

$$i\hbar\frac{d\rho_{ij}(t)}{dt} = \sum_k [(E_k\delta_{ik} + \langle i|\hat{V}(t)|k\rangle)\rho_{kj}(t)$$

$$- \rho_{ik}(t)(E_k\delta_{kj} + \langle k|\hat{V}(t)|j\rangle)], \tag{5.3.21}$$

and all the time dependence is shown. Equation (5.3.21) reduces to

$$i\hbar\frac{d\rho_{ij}(t)}{dt} = (E_i - E_j)\rho_{ij}(t) + \sum_k (\langle i|\hat{V}(t)|k\rangle\rho_{kj}(t)$$

$$- \rho_{ik}(t)\langle k|\hat{V}(t)|j\rangle). \tag{5.3.22}$$

We once again recognize the matrix multiplication and the presence of the commutator, so

$$i\hbar\frac{d\rho_{ij}(t)}{dt} = (E_i - E_j)\rho_{ij}(t) + \langle i|[\hat{V}(t), \hat{\rho}(t)]|j\rangle. \tag{5.3.23}$$

This equation provides a differential equation for each density matrix element. In fact, we have a coupled set of ordinary differential equations to solve. Equation (5.3.23) is the Liouville–von Neumann Equation for Hamiltonians with the structure of Eq. (5.3.18).

These last results are simplified in the Interaction Picture of Sec. 3.5, which "removes" the time dependence due to $\hat{H}_0$ from $|\psi(t)\rangle$. We define

$$|\tilde{\psi}(t)\rangle = e^{i\hat{H}_0t/\hbar}|\psi(t)\rangle, \tag{5.3.24}$$

as the ket in the Interaction Picture. We next develop the Interaction Picture operator based on $\hat{V}$. This requires us to return to $|\psi(t)\rangle$ and find two expressions for its time derivative. We start with Eq. (5.3.24) and find

$$i\hbar\frac{d|\psi(t)\rangle}{dt} = i\hbar\frac{d}{dt}(e^{-i\hat{H}_0t/\hbar}|\tilde{\psi}(t)\rangle)$$

$$= i\hbar(-i\hat{H}_0/\hbar)e^{-i\hat{H}_0t/\hbar}|\tilde{\psi}(t)\rangle + i\hbar e^{-i\hat{H}_0t/\hbar}\frac{d|\tilde{\psi}(t)\rangle}{dt}.$$

$$(5.3.25)$$

We also have the Time-Dependent Schrödinger Equation,

$$i\hbar\frac{d|\psi(t)\rangle}{dt} = [\hat{H}_0 + \hat{V}(t)]|\psi(t)\rangle. \qquad (5.3.26)$$

The right-hand sides of Eqs. (5.3.25) and (5.3.26) are equal and we see with Eq. (5.3.24)

$$\hat{H}_0|\psi(t)\rangle + i\hbar e^{-i\hat{H}_0t/\hbar}\frac{d|\tilde{\psi}(t)\rangle}{dt} = \hat{H}_0|\psi(t)\rangle + \hat{V}(t)|\psi(t)\rangle. \qquad (5.3.27)$$

We subtract the first terms on each side and again use Eq. (5.3.24) to find

$$i\hbar e^{-i\hat{H}_0t/\hbar}\frac{d|\tilde{\psi}(t)\rangle}{dt} = \hat{V}(t)|\psi(t)\rangle = \hat{V}(t)e^{-i\hat{H}_0t/\hbar}|\tilde{\psi}(t)\rangle, \qquad (5.3.28)$$

which we rewrite as

$$i\hbar\frac{d|\tilde{\psi}(t)\rangle}{dt} = e^{+i\hat{H}_0t/\hbar}\hat{V}(t)e^{-i\hat{H}_0t/\hbar}|\tilde{\psi}(t)\rangle = \hat{V}_I(t)|\tilde{\psi}(t)\rangle. \qquad (5.3.29)$$

Here, the last equality defines the potential operator in the Interaction Picture. Equation (5.3.29) shows explicitly that the time dependence of $|\tilde{\psi}(t)\rangle$ comes directly from the time dependence of $\hat{V}(t)$.

Finally, we derive the Liouville–von Neumann Equation for the density operator in the Interaction Picture $\hat{\rho}_I$ where

$$\hat{\rho}_I(t) = e^{i\hat{H}_0 t/\hbar} \hat{\rho}(t) e^{-i\hat{H}_0 t/\hbar}. \tag{5.3.30}$$

We start with Eq. (5.3.5) and use Eq. (5.3.18) to get

$$i\hbar\frac{d\hat{\rho}(t)}{dt} = \hat{H}_0\hat{\rho}(t) - \hat{\rho}(t)\hat{H}_0 + \hat{V}(t)\hat{\rho}(t) - \hat{\rho}(t)\hat{V}(t). \tag{5.3.31}$$

We next express $\hat{\rho}(t)$ through Eq. (5.3.30) and differentiate each term with respect to time to find

$$i\hbar\frac{d\hat{\rho}(t)}{dt} = i\hbar\left(-\frac{i}{\hbar}\hat{H}_0\right)e^{-i\hat{H}_0 t/\hbar}\hat{\rho}_I(t)e^{i\hat{H}_0 t/\hbar}$$

$$+ i\hbar e^{-i\hat{H}_0 t/\hbar}\frac{d\hat{\rho}_I(t)}{dt}e^{i\hat{H}_0 t/\hbar}$$

$$+ i\hbar e^{-i\hat{H}_0 t/\hbar}\hat{\rho}_I(t)\left(\frac{i}{\hbar}\hat{H}_0\right)e^{i\hat{H}_0 t/\hbar}. \tag{5.3.32}$$

The right-hand sides of Eqs. (5.3.31) and (5.3.32) are equal. Now the first and third terms of Eq. (5.3.32) are equal to the first two terms of Eq. (5.3.31). With their removal, we are left with

$$\hat{V}(t)\hat{\rho}(t) - \hat{\rho}(t)\hat{V}(t) = i\hbar e^{-i\hat{H}_0 t/\hbar}\frac{d\hat{\rho}_I(t)}{dt}e^{i\hat{H}_0 t/\hbar}. \tag{5.3.33}$$

This transforms into

$$i\hbar\frac{d\hat{\rho}_I(t)}{dt} = e^{i\hat{H}_0 t/\hbar}[\hat{V}(t)\hat{\rho}(t) - \hat{\rho}(t)\hat{V}(t)]e^{-i\hat{H}_0 t/\hbar}, \tag{5.3.34}$$

and with the insertion of more exponentials between the operators, we arrive at

$$i\hbar\frac{d\hat{\rho}_I(t)}{dt} = \hat{V}_I(t)\hat{\rho}_I(t) - \hat{\rho}_I(t)\hat{V}_I(t) = [\hat{V}_I(t), \hat{\rho}_I(t)]. \tag{5.3.35}$$

And this is the Liouville–von Neumann Equation for the density operator in the Interaction Picture. We solve this equation for a linearly-polarized electromagnetic field in the next section.

5.4. The Density Matrix Applied to a Two-Level System

We return to the time-dependent problem solved in Secs. 3.1 and 3.2. This problem is worked out in the context of the density matrix in Fayer (2001, Sec. 14.6). The following uses Chap. 3's notation, with the exception of the change from H_i to E_i. The complete Hamiltonian is

$$\hat{H}(t) = \hat{H}_0 + \hat{V}_c(t) = \hat{H}_0 + \hat{\mu} E_0 \cos(\omega t). \qquad (5.4.1)$$

Here $\hat{H}_0$ is independent of the time t, and we use the kets of Eq. (5.3.19). E_0 is the magnitude of the applied electric field that varies in time with angular frequency ω. The electric dipole moment operator is $\hat{\mu}$. We start with the transformation of $\hat{V}_c(t)$ into the Interaction Picture as in Eq. (5.3.29) and then use Eq. (5.3.35). This leads to equations for the matrix elements ρ_{Iij} that have the structure given in Eqs. (5.3.10) to (5.3.13), with $\hat{V}_I$ in place of $\hat{H}$. We first note that

$$\langle i|\hat{V}_c(t)|i\rangle = \langle i|\hat{\mu}|i\rangle E_0 \cos \omega t = 0, \qquad (5.4.2)$$

by symmetry, since the operator $\hat{\mu}$ is odd. Then for $i \neq j$,

$$V_{Iij} = \langle i|e^{i\hat{H}_0t/\hbar}\hat{V}_c e^{-i\hat{H}_0t/\hbar}|j\rangle = e^{iE_it/\hbar}e^{-iE_jt/\hbar}\langle i|\hat{\mu}|j\rangle E_0 \cos \omega t. \qquad (5.4.3)$$

As in Chap. 3, we define the angular frequencies

$$\tilde{\Omega} = \langle i|\hat{\mu}|j\rangle E_0/\hbar, \qquad (5.4.4)$$

and

$$\omega_0 = (E_2 - E_1)/\hbar. \qquad (5.4.5)$$

Now the matrix elements V_{Iij} are

$$[V_{Iij}/\hbar] = \begin{bmatrix} 0 & \tilde{\Omega} \cos \omega t e^{-i\omega_0 t} \\ \tilde{\Omega} \cos \omega t e^{+i\omega_0 t} & 0 \end{bmatrix}. \qquad (5.4.6)$$

The equations for the density matrix elements in the Interaction Picture are now:

$$i\frac{d\rho_{I11}}{dt} = \tilde{\Omega}\cos\omega t(e^{-i\omega_0 t}\rho_{I21} - e^{i\omega_0 t}\rho_{I12}), \qquad (5.4.7)$$

$$i\frac{d\rho_{I12}}{dt} = \tilde{\Omega}\cos\omega t e^{-i\omega_0 t}(\rho_{I22} - \rho_{I11}), \qquad (5.4.8)$$

$$i\frac{d\rho_{I21}}{dt} = \tilde{\Omega}\cos\omega t e^{i\omega_0 t}(\rho_{I11} - \rho_{I22}), \qquad (5.4.9)$$

$$i\frac{d\rho_{I22}}{dt} = \tilde{\Omega}\cos\omega t(e^{i\omega_0 t}\rho_{I12} - e^{-i\omega_0 t}\rho_{I21}). \qquad (5.4.10)$$

Once again, after setting

$$\cos\omega t = \frac{1}{2}(e^{i\omega t} + e^{-i\omega t}), \qquad (5.4.11)$$

we introduce the Rotating Wave Approximation (RWA) and drop terms with $(\omega + \omega_0)$ in the exponential. Our 4 coupled equations become:

$$i\frac{d\rho_{I11}}{dt} = \frac{\tilde{\Omega}}{2}(e^{i(\omega-\omega_0)t}\rho_{I21} - e^{-i(\omega-\omega_0)t}\rho_{I12}), \qquad (5.4.12)$$

$$i\frac{d\rho_{I12}}{dt} = \frac{\tilde{\Omega}}{2}e^{i(\omega-\omega_0)t}(\rho_{I22} - \rho_{I11}), \qquad (5.4.13)$$

$$i\frac{d\rho_{I21}}{dt} = \frac{\tilde{\Omega}}{2}e^{-i(\omega-\omega_0)t}(\rho_{I11} - \rho_{I22}), \qquad (5.4.14)$$

$$i\frac{d\rho_{I22}}{dt} = \frac{\tilde{\Omega}}{2}(e^{-i(\omega-\omega_0)t}\rho_{I12} - e^{i(\omega-\omega_0)t}\rho_{I21}). \qquad (5.4.15)$$

The complete solutions for Eqs. (5.4.12) to (5.4.15) appear without any derivation in Eq. (14.71) of Fayer (2001, Chap. 14). Ramble 5 of Sec. 5.10 contains the details of this solution, courtesy of Prof. Michael D. Fayer. A few preliminaries are useful before the solutions are stated here. The initial conditions are

$$\rho_{I11}(t = 0) = 1, \qquad (5.4.16)$$

$$\rho_{I12}(t = 0) = \rho_{I21}(t = 0) = \rho_{I22}(t = 0) = 0. \qquad (5.4.17)$$

This corresponds to starting the two-level system in the lower energy state. We also define

$$\delta\omega = \omega - \omega_0, \tag{5.4.18}$$

$$\Omega = [(\delta\omega)^2 + \tilde{\Omega}^2]^{1/2}. \tag{5.4.19}$$

Prof. Fayer's solutions follow in the present notation:

$$\rho_{I11}(t) = 1 - \left(\frac{\tilde{\Omega}}{\Omega}\right)^2 \sin^2(\Omega t/2), \tag{5.4.20}$$

$$\rho_{I12}(t) = \frac{\tilde{\Omega}}{\Omega^2}\left[\frac{i\Omega}{2}\sin\Omega t + \delta\omega\sin^2(\Omega t/2)\right]e^{i\delta\omega t}, \tag{5.4.21}$$

$$\rho_{I21}(t) = \frac{\tilde{\Omega}}{\Omega^2}\left[-\frac{i\Omega}{2}\sin\Omega t + \delta\omega\sin^2(\Omega t/2)\right]e^{-i\delta\omega t}, \tag{5.4.22}$$

$$\rho_{I22}(t) = \left(\frac{\tilde{\Omega}}{\Omega}\right)^2 \sin^2(\Omega t/2). \tag{5.4.23}$$

A straightforward derivation of these solutions presents a challenge. The occupation probabilities $\rho_{Iii}(t)$ appear in Eqs. (3.2.36) and (3.2.37), and the latter clearly agrees with Eq. (5.4.23). Equation (5.4.19) is used to turn Eq. (3.2.36) into Eq. (5.4.20). Thus, as an alternative to Sec. 5.10, we could start with the $\rho_{Iii}(t)$ and find $\rho_{I12}(t)$ and $\rho_{I21}(t)$ through Eqs. (5.4.13) and (5.4.14).

However, when $\delta\omega = 0$ we are at resonance and the differential equations for the $\rho_{Iij}(t)$ simplify,

$$i\frac{d\rho_{I11}}{dt} = \frac{\tilde{\Omega}}{2}(\rho_{I21} - \rho_{I12}), \tag{5.4.24}$$

$$i\frac{d\rho_{I12}}{dt} = \frac{\tilde{\Omega}}{2}(\rho_{I22} - \rho_{I11}), \tag{5.4.25}$$

$$i\frac{d\rho_{I21}}{dt} = \frac{\tilde{\Omega}}{2}(\rho_{I11} - \rho_{I22}), \tag{5.4.26}$$

$$i\frac{d\rho_{I22}}{dt} = \frac{\tilde{\Omega}}{2}(\rho_{I12} - \rho_{I21}). \tag{5.4.27}$$

These equations are easier to solve due to the lack of time-dependent coefficients on the right-hand sides. We start by taking the time derivative of Eq. (5.4.24),

$$i\frac{d^2\rho_{I11}}{dt^2} = \frac{\tilde{\Omega}}{2}\left(\frac{d}{dt}\rho_{I21} - \frac{d}{dt}\rho_{I12}\right), \tag{5.4.28}$$

and substitute for the time derivatives on the right-hand side through Eqs. (5.4.25) and (5.4.26).

This yields

$$i\frac{d^2\rho_{I11}}{dt^2} = \frac{\tilde{\Omega}}{2}\left(\frac{\tilde{\Omega}}{2i}\right)(\rho_{I11} - \rho_{I22}) - \frac{\tilde{\Omega}}{2}\left(\frac{\tilde{\Omega}}{2i}\right)(\rho_{I22} - \rho_{I11})$$

$$= \frac{\tilde{\Omega}^2}{2i}(\rho_{I11} - \rho_{I22}). \tag{5.4.29}$$

We now apply

$$\rho_{I11} + \rho_{I22} = 1, \tag{5.4.30}$$

and find

$$i\frac{d^2\rho_{I11}}{dt^2} = \frac{\tilde{\Omega}^2}{2i}(2\rho_{I11} - 1). \tag{5.4.31}$$

We rewrite this as

$$\frac{d^2\rho_{I11}}{dt^2} + \tilde{\Omega}^2\rho_{I11} - \frac{\tilde{\Omega}^2}{2} = 0, \tag{5.4.32}$$

which has the solution

$$\rho_{I11}(t) = A\cos\tilde{\Omega}t + B\sin\tilde{\Omega}t + C. \tag{5.4.33}$$

We substitute this into Eq. (5.4.32) to learn

$$\tilde{\Omega}^2 C - \frac{\tilde{\Omega}^2}{2} = 0, \tag{5.4.34}$$

and

$$C = \frac{1}{2}. \tag{5.4.35}$$

We assume the initial conditions given in Eqs. (5.4.16) and (5.4.17), and these lead to

$$\rho_{I11}(t = 0) = 1, \qquad (5.4.36)$$

and

$$\left.\frac{d\rho_{I11}}{dt}\right|_{t=0} = 0, \qquad (5.4.37)$$

by virtue of Eq. (5.4.24). When we apply these to Eq. (5.4.33) with Eq. (5.4.35), we find

$$A + \frac{1}{2} = 1, \qquad (5.4.38)$$

and

$$B = 0, \qquad (5.4.39)$$

thus,

$$\rho_{I11}(t) = \frac{1}{2}(\cos \tilde{\Omega}t + 1) = \frac{1}{2}(2\cos^2(\tilde{\Omega}t/2) - 1 + 1) = \cos^2(\tilde{\Omega}t/2). \qquad (5.4.40)$$

And by way of Eq. (5.4.30), the conservation of occupation probabilities, we have

$$\rho_{I22}(t) = \sin^2(\tilde{\Omega}t/2). \qquad (5.4.41)$$

These solutions for $\rho_{Iii}(t)$ at resonance agree with the solutions in Eqs. (5.4.20) and (5.4.23), since at resonance

$$\tilde{\Omega} = \Omega. \qquad (5.4.42)$$

We are left with Eqs. (5.4.25) and (5.4.26). We substitute the above results for $\rho_{Iii}(t)$ into Eq. (5.4.25), as suggested after Eq. (5.4.23), to find,

$$i\frac{d}{dt}\rho_{I12}(t) = \frac{\tilde{\Omega}}{2}[\sin^2(\tilde{\Omega}t/2) - \cos^2(\tilde{\Omega}t/2)] = -\frac{\tilde{\Omega}}{2}\cos \tilde{\Omega}t. \qquad (5.4.43)$$

This differential equation has the solution

$$\rho_{I12}(t) = \frac{i}{2} \sin \tilde{\Omega} t, \tag{5.4.44}$$

and

$$\rho_{I21}(t) = (\rho_{I12}(t))^* = -\frac{i}{2} \sin \tilde{\Omega} t. \tag{5.4.45}$$

These last two results agree with the solutions in Eqs. (5.4.21) and (5.4.22) when they are evaluated at resonance. In addition, the result in Eq. (5.4.45) satisfies Eq. (5.4.26).

Thus, we have found the density matrix elements in the Interaction Picture for a time-varying electric field. This supplements the results of Sec. 3.2. If we turn off the electric field at time $t = t'$, then the $\rho_{Iij}(t)$ stay at $\rho_{Iij}(t')$. Physically, we expect the $\rho_{Iii}(t)$ to approach their thermal equilibrium values as the time increases

$$\rho_{I22}/\rho_{I11} = e^{-\hbar\omega_0/k_B T}, \tag{5.4.46}$$

with k_B Boltzmann's constant and the temperature T in kelvin. In addition, we expect the off-diagonal density matrix elements to go to zero. This behavior does occur when we introduce relaxation terms into the equations for the density matrix elements. This is our next task.

5.5. The Introduction of Relaxation into the Density Matrix

Generally, our two-level system interacts with its environment. This may consist of other two-level systems, electrons, nucleons, additional atoms and/or spins. In many cases, the environment may be viewed as a thermal reservoir or bath that exchanges energy with the two-level system. In addition, we may consider spontaneous emission to result from the interaction of the two-level system with the vacuum (Cohen-Tannoudji *et al.*, 1992, Sec. III-C). Numerous approaches have been developed to tackle the coupling between a system of interest and its environment. If we make the assumption that the system of interest does not affect its environment, then the reduced density operator and the reduced density matrix are useful.

Shore (1990, Sec. 6.4) and Blum (1996, Chap. 8) explain these ideas. Both treatments discuss master equations and their use in describing physical systems. In addition, Dubbers and Stöckmann (2013, Chap. 22) work through a perturbative and a stochastic approach to system-environment interactions, that is, relaxation.

We take a more phenomenological approach and add relaxation terms to the Liouville–von Neumann Equations. We work in the Schrödinger Picture. Blum (1996, Chap. 8) provides formal derivations for this method. We start with Eqs. (5.3.10) to (5.3.13) for the density matrix elements and rewrite them here for convenience:

$$i\hbar\frac{d\rho_{11}(t)}{dt} = H_{12}\rho_{21} - H_{21}\rho_{12}, \tag{5.5.1}$$

$$i\hbar\frac{d\rho_{12}(t)}{dt} = (H_{11} - H_{22})\rho_{12} + H_{12}(\rho_{22} - \rho_{11}), \tag{5.5.2}$$

$$i\hbar\frac{d\rho_{21}(t)}{dt} = (H_{22} - H_{11})\rho_{21} + H_{21}(\rho_{11} - \rho_{22}), \tag{5.5.3}$$

$$i\hbar\frac{d\rho_{22}(t)}{dt} = H_{21}\rho_{12} - H_{12}\rho_{21}. \tag{5.5.4}$$

Now we need to introduce relaxation. We do this by considering Eqs. (5.5.1) to (5.5.4) when the external field is absent. Thus, we desire the ρ_{ii}, which are the occupation probabilities, over time to approach their equilibrium values and the off-diagonal ρ_{ij}, which are the coherences, to go to zero with time. Next, we see what terms may be added to these equations to achieve these limits. The Hamiltonian $\hat{H}$ acts on the states of our two-level system and we continue to let

$$\hat{H} = \hat{H}_0 + \hat{V}_c. \tag{5.5.5}$$

We use the base kets $|i\rangle$ that are the orthonormal eigenkets of $\hat{H}_0$,

$$\hat{H}_0|i\rangle = E_i|i\rangle. \tag{5.5.6}$$

Equations (5.5.1) to (5.5.4) with $\hat{V}_c = 0$ have $H_{12} = H_{21} = 0$ and become:

$$i\hbar\frac{d\rho_{11}(t)}{dt} = 0, \tag{5.5.7}$$

$$i\hbar \frac{d\rho_{12}(t)}{dt} = (E_1 - E_2)\rho_{12}(t), \qquad (5.5.8)$$

$$i\hbar \frac{d\rho_{21}(t)}{dt} = (E_2 - E_1)\rho_{21}(t), \qquad (5.5.9)$$

$$i\hbar \frac{d\rho_{22}(t)}{dt} = 0. \qquad (5.5.10)$$

Here we repeatedly use

$$\langle i|\hat{H}_0|j\rangle = E_j\langle i|j\rangle = 0, \qquad (5.5.11)$$

for $i \neq j$. We note that Eqs. (5.5.7) to (5.5.10) say $\rho_{ii}(t)$ is a constant and the off-diagonal $\rho_{ij}(t)$ are not zero. Hence, we need to include additional terms and we start with the equations for $\rho_{ii}(t)$.

The diagonal density matrix elements will approach their equilibrium values ρ_{ii}^{eq} if we assume

$$i\hbar \frac{d\rho_{ii}(t)}{dt} = \frac{i\hbar}{T_1}(\rho_{ii}^{eq} - \rho_{ii}(t)), \qquad (5.5.12)$$

where T_1 is a relaxation time. This provides the simplest representation of the interaction of the two-level system with its environment. We find the solution to Eq. (5.5.12) to be

$$\rho_{ii}(t) = \rho_{ii}^{eq} + [\rho_{ii}(t_0) - \rho_{ii}^{eq}]e^{-(t-t_0)/T_1}, \qquad (5.5.13)$$

for $t \geq t_0$ with t_0 the time $\hat{V}_c$ is set to zero. We see that as the time approaches infinity,

$$\rho_{ii}(t \to \infty) \to \rho_{ii}^{eq}, \qquad (5.5.14)$$

as desired.

We next introduce a relaxation time T_2 for the off-diagonal density matrix elements. After $\hat{V}_c$ is set to zero, we want the coherences, or the off-diagonal $\rho_{ij}(t)$, to evolve in time towards zero. We expect this to happen independently of the changes in the occupation probabilities within the present model of relaxation, so we bring in T_2.

We take $i \neq j$ and write

$$i\hbar \frac{d\rho_{ij}(t)}{dt} = (E_i - E_j)\rho_{ij}(t) - \frac{i\hbar}{T_2}\rho_{ij}(t), \qquad (5.5.15)$$

or to be specific

$$\frac{d\rho_{12}(t)}{dt} = -\frac{\omega_0}{i}\rho_{12}(t) - \frac{1}{T_2}\rho_{12}(t) = \left(i\omega_0 - \frac{1}{T_2}\right)\rho_{12}(t), \quad (5.5.16)$$

with

$$\omega_0 = (E_2 - E_1)/\hbar. \qquad (5.5.17)$$

The solution to Eq. (5.5.16) is

$$\rho_{12}(t) = \rho_{12}(t_0)e^{(i\omega_0 - \frac{1}{T_2})(t-t_0)}, \qquad (5.5.18)$$

and we see exponential damping along with an oscillatory behavior. We note that this yields

$$\rho_{12}(t \to \infty) \to 0. \qquad (5.5.19)$$

A similar solution results for $\rho_{21}(t)$.

Thus, we have introduced two time constants to characterize the relaxation of a two-level system to equilibrium when the coupling interaction $\hat{V}_c$ is turned off. We assume the environment is providing the energy needed for any transitions from $|1\rangle$ to $|2\rangle$ required for $\rho_{11}(t)$ to approach ρ_{11}^{eq}. T_1 is often called the longitudinal relaxation time and in solids represents the interaction between the lattice and the two-level system. T_2 is the transverse relaxation time and it represents the interaction of the two-level system with similar systems. Spin-spin interactions are an example. Slichter (2010) and Skomski (2008) provide details on the T_i. Both T_1 and T_2 may depend on the details of the system and its environment. In addition, these relaxation times may be functions of the temperature and whatever fields are present.

We also assume these relaxation terms belong in the Liouville–von Neumann Equations when $\hat{V}_c$ is on. We further assume that both diagonal density matrix elements have the same relaxation time T_1

and that both off-diagonal density matrix elements have the same T_2. Our Liouville–von Neumann equations are now

$$i\hbar\frac{d\rho_{11}(t)}{dt} = H_{12}\rho_{21}(t) - H_{21}\rho_{12}(t) + \frac{i\hbar}{T_1}[\rho_{11}^{eq} - \rho_{11}(t)], \qquad (5.5.20)$$

$$i\hbar\frac{d\rho_{12}(t)}{dt} = (H_{11} - H_{22})\rho_{12}(t) + H_{12}(\rho_{22}(t) - \rho_{11}(t)) - \frac{i\hbar}{T_2}\rho_{12}(t),$$
$$(5.5.21)$$

$$i\hbar\frac{d\rho_{21}(t)}{dt} = (H_{22} - H_{11})\rho_{21}(t) + H_{21}(\rho_{11}(t) - \rho_{22}(t)) - \frac{i\hbar}{T_2}\rho_{21}(t),$$
$$(5.5.22)$$

$$i\hbar\frac{d\rho_{22}(t)}{dt} = H_{21}\rho_{12}(t) - H_{12}\rho_{21}(t) + \frac{i\hbar}{T_1}[\rho_{22}^{eq} - \rho_{22}(t)]. \qquad (5.5.23)$$

These equations include the coupling interaction $\hat{V}_c$, so the matrix elements of the Hamiltonian may well have time dependences. It is apparent that the introduction of relaxation makes the Liouville–von Neumann equations harder to solve in closed-form. The next section presents a rare example of a closed-form solution.

5.6. A Solvable Two-Level System with Relaxation Included

Equations (5.5.1) to (5.5.4) are sufficiently challenging by themselves. The degree of difficulty is increased by the introduction of the phenomenological relaxation terms in Eqs. (5.5.20) to (5.5.23). Still, it is worthwhile to examine the role of relaxation while the coupling interaction $\hat{V}_c$ is active, even though this requires many assumptions in order to get a closed-form solution. We first discuss a solution of the Liouville–von Neumann Equations for an example from Bittner (2010, Sec. 6.4.3). Section 5.8 contains an additional example in the language of the Bloch Equations, which are based on combinations of the density matrix elements. The present section ends with an example of relaxation in a two-level system within a quantum dot.

Our treatment departs slightly from the cases Bittner details, in that we allow the ρ_{ii}^{eq} to be non-zero. We stay in the Schrödinger Picture. We assume the coupling interaction $\hat{V}_c$ is a real constant V, so

$$H_{12} = H_{21} = V. \tag{5.6.1}$$

We further assume the relaxation time for the diagonal density matrix elements is T_1 and the relaxation time for the off-diagonal density matrix elements is T_2. We also assume

$$H_{11} = H_{22}. \tag{5.6.2}$$

With these assumptions, the Liouville–von Neumann Equations, Eqs. (5.5.20) to (5.5.23), become:

$$\frac{d\rho_{11}(t)}{dt} = \frac{V}{i\hbar}[\rho_{21}(t) - \rho_{12}(t)] + \frac{1}{T_1}[\rho_{11}^{eq} - \rho_{11}(t)], \tag{5.6.3}$$

$$\frac{d\rho_{12}(t)}{dt} = \frac{V}{i\hbar}[\rho_{22}(t) - \rho_{11}(t)] - \frac{1}{T_2}\rho_{12}(t), \tag{5.6.4}$$

$$\frac{d\rho_{21}(t)}{dt} = \frac{V}{i\hbar}[\rho_{11}(t) - \rho_{22}(t)] - \frac{1}{T_2}\rho_{21}(t), \tag{5.6.5}$$

$$\frac{d\rho_{22}(t)}{dt} = \frac{V}{i\hbar}[\rho_{12}(t) - \rho_{21}(t)] + \frac{1}{T_1}[\rho_{22}^{eq} - \rho_{22}(t)]. \tag{5.6.6}$$

We still have

$$\rho_{11}(t) + \rho_{22}(t) = 1, \tag{5.6.7}$$

and with Eqs. (1.4.10) and (5.6.2), the equilibrium values are

$$\rho_{ii}^{eq} = 0.5. \tag{5.6.8}$$

We first take the time derivative of Eq. (5.6.3),

$$\frac{d^2\rho_{11}(t)}{dt^2} = \frac{V}{i\hbar}\left[\frac{d\rho_{21}(t)}{dt} - \frac{d\rho_{12}(t)}{dt}\right] - \frac{1}{T_1}\frac{d\rho_{11}(t)}{dt}. \tag{5.6.9}$$

We next combine Eqs. (5.6.4) and (5.6.5) to find

$$\frac{d\rho_{21}(t)}{dt} - \frac{d\rho_{12}(t)}{dt} = -\frac{2V}{i\hbar}[\rho_{22}(t) - \rho_{11}(t)] - \frac{1}{T_2}[\rho_{21}(t) - \rho_{12}(t)], \tag{5.6.10}$$

and substitute this into Eq. (5.6.9),

$$\frac{d^2\rho_{11}(t)}{dt^2} = \frac{2V^2}{\hbar^2}[\rho_{22}(t) - \rho_{11}(t)]$$

$$- \frac{V}{i\hbar T_2}[\rho_{21}(t) - \rho_{12}(t)] - \frac{1}{T_1}\frac{d\rho_{11}(t)}{dt}. \quad (5.6.11)$$

We use Eq. (5.6.3) to replace $[\rho_{21}(t) - \rho_{12}(t)]$, so

$$\frac{d^2\rho_{11}(t)}{dt^2} = \frac{2V^2}{\hbar^2}[\rho_{22}(t) - \rho_{11}(t)]$$

$$- \frac{1}{T_2}\left[\frac{d\rho_{11}(t)}{dt} - \frac{1}{T_1}[\rho_{11}^{eq} - \rho_{11}(t)]\right]$$

$$- \frac{1}{T_1}\frac{d\rho_{11}(t)}{dt}. \quad (5.6.12)$$

We finally use Eq. (5.6.7) to eliminate $\rho_{22}(t)$,

$$\frac{d^2\rho_{11}(t)}{dt^2} = \frac{2V^2}{\hbar^2}[1 - 2\rho_{11}(t)] - \left(\frac{1}{T_1} + \frac{1}{T_2}\right)\frac{d\rho_{11}(t)}{dt}$$

$$+ \frac{1}{T_1 T_2}[\rho_{11}^{eq} - \rho_{11}(t)]. \quad (5.6.13)$$

This equation is rewritten as

$$\frac{d^2\rho_{11}(t)}{dt^2} + \left(\frac{1}{T_1} + \frac{1}{T_2}\right)\frac{d\rho_{11}(t)}{dt} + \left(\frac{4V^2}{\hbar^2} + \frac{1}{T_1 T_2}\right)\rho_{11}(t)$$

$$= \frac{2V^2}{\hbar^2} + \frac{\rho_{11}^{eq}}{T_1 T_2}, \quad (5.6.14)$$

which resembles the equation for a damped oscillator.

For ease, Eq. (5.6.14) is written in terms of the constants A, B and C that are defined by direct comparison,

$$\frac{d^2\rho_{11}(t)}{dt^2} + A\frac{d\rho_{11}(t)}{dt} + B\rho_{11}(t) = C. \quad (5.6.15)$$

The homogeneous, second-order, ordinary differential equation taken from Eq. (5.6.15) is solved by exponentials in time such as

$$ae^{\gamma t}, \quad (5.6.16)$$

and the γ come from substituting Eq. (5.6.16) into the homogeneous version of Eq. (5.6.15). This leads to

$$\gamma^2 + A\gamma + B = 0, \tag{5.6.17}$$

hence,

$$\gamma_\pm = -\frac{A}{2} \pm \frac{1}{2}\sqrt{A^2 - 4B}. \tag{5.6.18}$$

A glance at Eq. (5.6.14) tells us that both A and B are positive, so $\gamma_\pm$ have negative real parts since the square root in Eq. (5.6.18) is less than A. This means $\rho_{11}(t)$ is bounded as the time increases. And if $A^2 < 4B$, then $\gamma_\pm$ have imaginary parts and $\rho_{11}(t)$ oscillates in time.

The complete solution to Eq. (5.6.15) is

$$\rho_{11}(t) = a_- e^{\gamma_- t} + a_+ e^{\gamma_+ t} + (C/B), \tag{5.6.19}$$

as is verified by substitution into Eq. (5.6.15). The constants a_- and a_+ are set by the initial conditions. We take

$$\rho_{11}(0) = 1, \tag{5.6.20}$$

$$\rho_{12}(0) = \rho_{21}(0) = 0, \tag{5.6.21}$$

$$\rho_{22}(0) = 0. \tag{5.6.22}$$

Equation (5.6.19) now becomes

$$\rho_{11}(0) = a_- + a_+ + (C/B) = 1, \tag{5.6.23}$$

while the other initial conditions in Eq. (5.6.21) take Eq. (5.6.3) to

$$\frac{d\rho_{11}(t=0)}{dt} = \frac{1}{T_1}(\rho_{11}^{eq} - 1) = a_-\gamma_- + a_+\gamma_+, \tag{5.6.24}$$

where we have used the time derivative of Eq. (5.6.19) also. We need to solve Eqs. (5.6.23) and (5.6.24) for a_- and a_+. In doing so, we find,

$$a_- = \frac{\gamma_+ \left(1 - \frac{C}{B}\right) - \frac{1}{T_1}\left(\rho_{11}^{eq} - 1\right)}{\gamma_+ - \gamma_-}, \tag{5.6.25}$$

and

$$a_+ = \frac{-\gamma_- \left(1 - \frac{C}{B}\right) + \frac{1}{T_1}\left(\rho_{11}^{eq} - 1\right)}{\gamma_+ - \gamma_-}. \tag{5.6.26}$$

This gives us $\rho_{11}(t)$ and by Eq. (5.6.7), $\rho_{22}(t)$. We then substitute these functions into Eqs. (5.6.4) and (5.6.5) and determine the off-diagonal density matrix elements.

We look now at the large-time behavior of the solution for $\rho_{11}(t)$. The real parts of γ_- and γ_+ are negative, so the exponential terms go to zero at large times. Hence,

$$\rho_{11}(t \to \infty) \to \frac{C}{B} = \frac{\frac{2V^2}{\hbar^2} + \frac{\rho_{11}^{eq}}{T_1 T_2}}{\frac{4V^2}{\hbar^2} + \frac{1}{T_1 T_2}}. \tag{5.6.27}$$

We have assumed the diagonal matrix elements of the Hamiltonian are equal per Eq. (5.6.2). This leads to Eq. (5.6.8) and $\rho_{11}^{eq} = 0.5$. When we substitute this into Eq. (5.6.27), we see

$$\rho_{11}(t \to \infty) \to \frac{C}{B} = \frac{1}{2} = \rho_{11}^{eq}, \tag{5.6.28}$$

and we have consistency.

Figures 5.1 and 5.2 show $\rho_{11}(t)$ based on the above solution and the initial conditions of Eqs. (5.6.20) to (5.6.22). The time is in terms of T_1, the relaxation time for the diagonal matrix elements of the density matrix. Two additional parameters are needed. The first is the strength of the coupling interaction

$$\frac{V}{\hbar} = \frac{K}{T_1}, \tag{5.6.29}$$

with the dimensionless parameter K. The second parameter Δ is also dimensionless and provides T_2, the relaxation time for the off-diagonal matrix elements of the density matrix, through

$$T_2 = \Delta T_1. \tag{5.6.30}$$

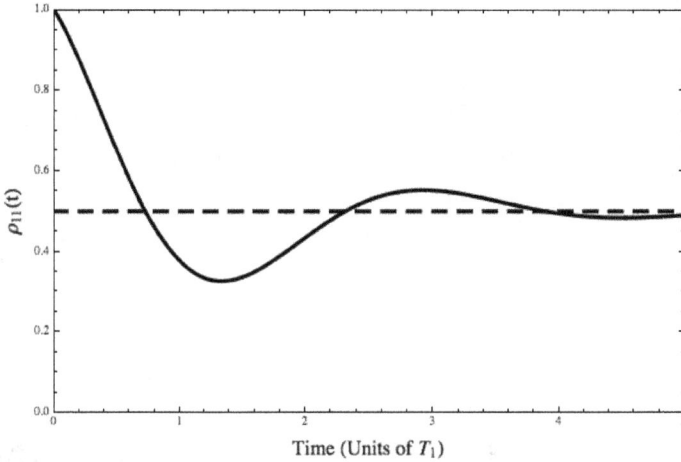

Figure 5.1. The diagonal density matrix element $\rho_{11}(t)$ (solid line) based on the solution of Eq. (5.6.14). The parameters are $K = 1.0$ and $\Delta = 0.50$. The dashed line is the equilibrium value $= 0.5$ for this example.

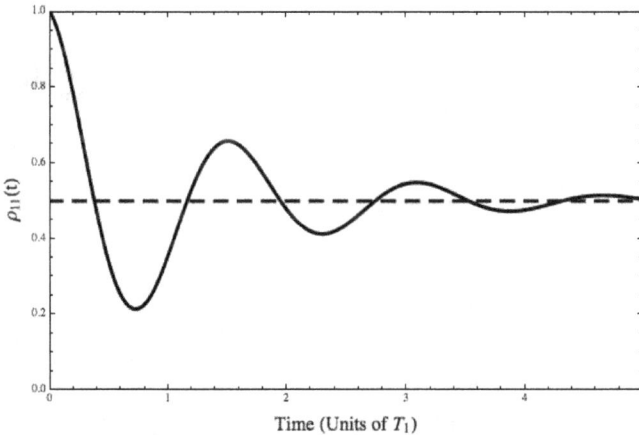

Figure 5.2. The diagonal density matrix element $\rho_{11}(t)$ (solid line) based on the solution of Eq. (5.6.14). The parameters are $K = 2.0$ and $\Delta = 0.50$. The dashed line is the equilibrium value $= 0.5$ for this example.

Figure 5.1 uses $K = 1.0$ and $\Delta = 0.50$. $\rho_{11}(t)$ drops from its initial value of 1.0 and oscillates about the equilibrium value shown by the dashed line. When K is increased to 2.0, as in Fig. 5.2, the initial decrease is faster and the oscillations are larger. These figures illustrate how the introduction of relaxation affects the Rabi oscillations of Chap. 3.

We have assumed $H_{11} = H_{22}$. If we remove this assumption, we find coupled equations due to a $\rho_{12} + \rho_{21}$ term when we develop the equation for $\rho_{11}(t)$. Bittner (2010, pp. 177–178) shows how Laplace transforms yield the time constant for $\rho_{11}(t)$ in this case.

Two-level systems in quantum dots provide examples of Rabi oscillations with relaxation. Figure 5.3 (Zrenner *et al.*, 2002) shows the damping of Rabi oscillations in the photocurrent for an illuminated indium gallium arsenide, InGaAs, quantum dot embedded in a photodiode. Quantum dots are nanostructures made from semiconductors such as InGaAs and gallium arsenide, GaAs, and are discussed in texts by Hanson (2008) and Chakraborty (1999). The horizontal axis in Fig. 5.3 relates to the Pulse Area Theorem of Sec. 4.5. Quantum dots are many-body systems with a plethora of possible relaxation mechanisms, hence, it is challenging to explain

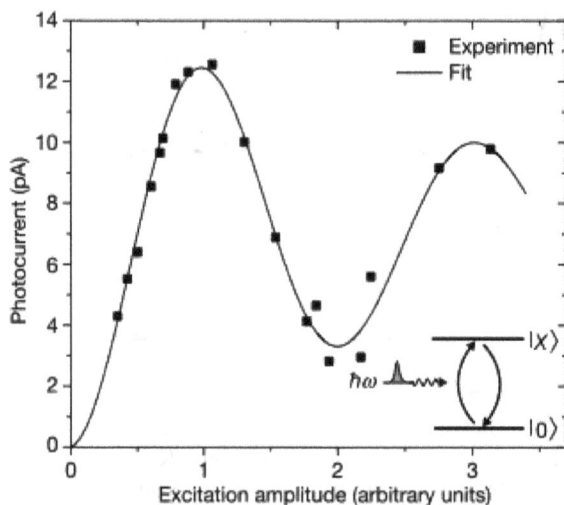

Figure 5.3. Damped Rabi oscillations from a quantum dot experiment by Zrenner *et al.* (2002, Fig. 3). The insert is a schematic of the transitions and is explained in the reference. This figure is used with the permission of the Nature Publishing Group. Reprinted by permission from Macmillan Publishers Ltd: *Nature* **418**, 612 (2002).

the sources of the damping seen in Fig. 5.3. Brandi *et al.* (2005) used a density matrix approach and provide a possible explanation of the data in Fig. 5.3 and the additional data from the same group in Beham *et al.* (2003). These experiments are done at low temperature so that the thermal energy, $k_B T$, is less than the energy level separation. Nadj-Perge *et al.* (2010) also report damped Rabi oscillations with quantum dots formed by a series of gate electrodes allied with an indium arsenide nanowire. Their experiments are done at 300 mK. These quantum dot experiments need low temperatures in contrast with the NV^- work discussed in Sec. 4.4. The NV^- defect has unusually slow spin relaxation which permits room temperature results.

Section 5.7 takes us from the matrix elements of the density matrix to the Bloch Equations.

5.7. The Bloch Equations

Our goal is to develop the Bloch Equations. We first restate the Liouville–von Neumann Equations with relaxation in terms of three coupled equations for a spin-$\frac{1}{2}$ system. These are related to the three independent functions for the 2×2 density matrix ρ discussed in Sec. 5.2. We find we may replace the ρ_{ij} with the functions U_1, U_2 and U_3, which we need to define. We then connect these $\{U_i\}$ to the average values of the components of the spin angular momentum. Finally, we derive time-dependent ordinary differential equations for these averages, and this leads us to the Bloch Equations.

In Sec. 5.2 we state that there are three independent functions for the 2×2 density matrix ρ. The ρ_{ii} are the occupation probabilities, so they are real. In addition, the trace is

$$\text{Tr}(\rho) = \rho_{11} + \rho_{22} = 1, \tag{5.7.1}$$

so only one ρ_{ii} is needed to determine both. This gives us our first independent function. The definitions of Eq. (5.2.8) and (5.2.9) for $i \neq j$ yield

$$\rho_{ij} = (\rho_{ji})^*. \tag{5.7.2}$$

Thus, it suffices to take ρ_{11} and the real and imaginary parts of an off-diagonal ρ_{ij}, for example, ρ_{21}, for the three independent functions. We define

$$\rho_{21} = \rho_{21}^R + i\rho_{21}^I. \tag{5.7.3}$$

We now develop useful combinations of the three independent functions. We start with

$$\rho_{11} - \rho_{22} = \rho_{11} - (1 - \rho_{11}) = 2\rho_{11} - 1. \tag{5.7.4}$$

Next,

$$\rho_{21} - \rho_{12} = \rho_{21} - (\rho_{21})^* = \rho_{21}^R + i\rho_{21}^I - (\rho_{21}^R - i\rho_{21}^I) = 2i\rho_{21}^I, \tag{5.7.5}$$

hence,

$$i(\rho_{21} - \rho_{12}) = -2\rho_{21}^I. \tag{5.7.6}$$

Similarly,

$$\rho_{21} + \rho_{12} = \rho_{21} + (\rho_{21})^* = 2\rho_{21}^R. \tag{5.7.7}$$

We define the useful combinations:

$$U_1 = \rho_{21} + \rho_{12}, \tag{5.7.8}$$

$$U_2 = i(\rho_{21} - \rho_{12}), \tag{5.7.9}$$

$$U_3 = \rho_{22} - \rho_{11}. \tag{5.7.10}$$

And these three U_i specify the three independent functions needed to completely specify the four density matrix elements for the 2×2 density matrix.

Our next task is to combine the Liouville–von Neumann Equations with relaxation from Eqs. (5.5.20) to (5.5.23) into three equations for the three U_i. We start with the equation for U_1,

$$i\hbar\frac{d(\rho_{21} + \rho_{12})}{dt} = (H_{22} - H_{11})(\rho_{21} - \rho_{12}) + (H_{12} - H_{21})(\rho_{22} - \rho_{11})$$

$$- \frac{i\hbar}{T_2}(\rho_{21} + \rho_{12}), \qquad (5.7.11)$$

hence,

$$i\hbar\frac{dU_1}{dt} = \frac{1}{i}(H_{22} - H_{11})U_2 + (H_{12} - H_{21})U_3 - \frac{i\hbar}{T_2}U_1. \qquad (5.7.12)$$

Next, for U_2,

$$i\hbar\frac{d(\rho_{21} - \rho_{12})}{dt} = (H_{22} - H_{11})(\rho_{21} + \rho_{12}) - (H_{12} + H_{21})(\rho_{22} - \rho_{11})$$

$$- \frac{i\hbar}{T_2}(\rho_{21} - \rho_{12}), \qquad (5.7.13)$$

or

$$\hbar\frac{dU_2}{dt} = (H_{22} - H_{11})U_1 - (H_{12} + H_{21})U_3 - \frac{\hbar}{T_2}U_2. \qquad (5.7.14)$$

Finally, for U_3,

$$i\hbar\frac{d(\rho_{22} - \rho_{11})}{dt} = 2H_{21}\rho_{12} - 2H_{12}\rho_{21} + \frac{i\hbar}{T_1}(\rho_{22}^{eq} - \rho_{11}^{eq})$$

$$- \frac{i\hbar}{T_1}(\rho_{22} - \rho_{11}). \qquad (5.7.15)$$

We add and subtract combinations of $H_{21}\rho_{21}$ and $H_{12}\rho_{12}$, which leads to

$$i\hbar\frac{dU_3}{dt} = (H_{21} - H_{12})(\rho_{21} + \rho_{12}) - (H_{21} + H_{12})(\rho_{21} - \rho_{12})$$

$$+ \frac{i\hbar}{T_1}(U_3^{eq} - U_3), \qquad (5.7.16)$$

and

$$i\hbar\frac{dU_3}{dt} = (H_{21} - H_{12})U_1 - \frac{1}{i}(H_{21} + H_{12})U_2 + \frac{i\hbar}{T_1}(U_3^{eq} - U_3).$$

$$(5.7.17)$$

Thus, Eqs. (5.7.12), (5.7.14) and (5.7.17) provide three coupled equations for U_1, U_2 and U_3. These, in turn, yield the density matrix elements for our two-level system.

We take a detour to bring in the spin angular momentum. First, we need the 2×2 density matrix ρ for our spin-$\frac{1}{2}$ two-level system in terms of the 2×2 identity matrix I and the three Pauli matrices $\{\sigma_i\}$, which are defined in Sec. 1.3. This requires four coefficients and we quickly find that one coefficient is easily found. Schatz and Ratner (2002, p. 268) also discuss the following expansion. We write

$$\rho = c_0 I + c_x \sigma_x + c_y \sigma_y + c_z \sigma_z, \tag{5.7.18}$$

and we determine the coefficients in steps with the aid of the trace Tr. To do this, we use

$$\text{Tr}(\rho I) = \text{Tr}(\rho) = 1, \tag{5.7.19}$$

in accord with Eq. (5.7.1) and

$$\text{Tr}(\sigma_i I) = \text{Tr}(\sigma_i) = 0. \tag{5.7.20}$$

This leaves us with

$$\text{Tr}(\rho I) = c_0 \text{Tr}(II) = c_0 \text{Tr}(I) = 2c_0, \tag{5.7.21}$$

and with Eq. (5.7.19) we find

$$c_0 = 0.5. \tag{5.7.22}$$

The other c_i follow from calculating the average value of each σ_i, $\langle \sigma_i \rangle$, and from

$$\text{Tr}(\sigma_i \sigma_j) = 0, \tag{5.7.23}$$

for $i \neq j$. Hence, we use Eq. (5.7.18) to find

$$\langle \sigma_i \rangle = \text{Tr}(\rho \sigma_i) = c_i \text{Tr}(\sigma_i \sigma_i) = c_i \text{Tr}(I) = 2c_i. \tag{5.7.24}$$

Equations (5.7.22) and (5.7.24) lead us to

$$\rho = \frac{1}{2} + \frac{1}{2}(\langle \sigma_x \rangle \sigma_x + \langle \sigma_y \rangle \sigma_y + \langle \sigma_z \rangle \sigma_z), \tag{5.7.25}$$

which we write in the form of a 2×2 matrix

$$\begin{bmatrix} \rho_{11} & \rho_{12} \\ \rho_{21} & \rho_{22} \end{bmatrix} = \begin{bmatrix} \frac{1}{2}(1 + \langle \sigma_z \rangle) & \frac{1}{2}(\langle \sigma_x \rangle - i\langle \sigma_y \rangle) \\ \frac{1}{2}(\langle \sigma_x \rangle + i\langle \sigma_y \rangle) & \frac{1}{2}(1 - \langle \sigma_z \rangle) \end{bmatrix}. \tag{5.7.26}$$

We next combine the matrix elements of the density matrix ρ in order to return to the $\{U_i\}$. First,

$$\rho_{21} + \rho_{12} = \langle \sigma_x \rangle, \tag{5.7.27}$$

$$\rho_{21} - \rho_{12} = i\langle \sigma_y \rangle, \tag{5.7.28}$$

$$\rho_{22} - \rho_{11} = -\langle \sigma_z \rangle. \tag{5.7.29}$$

We use Eqs. (5.7.8) to (5.7.10) and we note that the average values of the components of the spin angular momentum $\hat{\vec{S}}$ are $\hbar/2$ times the $\langle \sigma_i \rangle$. We put all this together and find

$$\langle \hat{S}_x \rangle = \frac{\hbar}{2} U_1, \tag{5.7.30}$$

$$\langle \hat{S}_y \rangle = -\frac{\hbar}{2} U_2, \tag{5.7.31}$$

$$\langle \hat{S}_z \rangle = -\frac{\hbar}{2} U_3. \tag{5.7.32}$$

Hence, we may express Eqs. (5.7.12), (5.7.14) and (5.7.17) in terms of the average values of the spin angular momentum. When this is done in the context of magnetic resonance studies, the Bloch Equations emerge.

Please beware! Some authors define U_2 and U_3 as the negative of the definitions used here. The present treatment has with the Hamiltonian $\hat{H}$

$$\Delta E = H_{22} - H_{11} \geq 0, \tag{5.7.33}$$

with $|1\rangle$ the ground state and $|2\rangle$ the excited state.

We end this section with the differential equations for the $\langle \hat{S}_i \rangle$. We use (5.7.12), (5.7.14) and (5.7.17) along with Eqs. (5.7.30) to

(5.7.33). First, for $\langle \hat{S}_x \rangle$,

$$i\frac{d\langle \hat{S}_x \rangle}{dt} = -\frac{i\Delta E}{\hbar}(-\langle \hat{S}_y \rangle) - \frac{(H_{12} - H_{21})}{\hbar}\langle \hat{S}_z \rangle - \frac{i}{T_2}\langle \hat{S}_x \rangle,$$

(5.7.34)

and

$$\frac{d\langle \hat{S}_x \rangle}{dt} = \frac{\Delta E}{\hbar}\langle \hat{S}_y \rangle - \frac{(H_{12} - H_{21})}{i\hbar}\langle \hat{S}_z \rangle - \frac{1}{T_2}\langle \hat{S}_x \rangle.$$

(5.7.35)

Next, for $\langle \hat{S}_y \rangle$,

$$-\frac{d\langle \hat{S}_y \rangle}{dt} = \frac{\Delta E}{\hbar}\langle \hat{S}_x \rangle + \frac{(H_{12} + H_{21})}{\hbar}\langle \hat{S}_z \rangle + \frac{1}{T_2}\langle \hat{S}_y \rangle,$$

(5.7.36)

and

$$\frac{d\langle \hat{S}_y \rangle}{dt} = -\frac{\Delta E}{\hbar}\langle \hat{S}_x \rangle - \frac{(H_{12} + H_{21})}{\hbar}\langle \hat{S}_z \rangle - \frac{1}{T_2}\langle \hat{S}_y \rangle.$$

(5.7.37)

Finally, for $\langle \hat{S}_z \rangle$,

$$-i\frac{d\langle \hat{S}_z \rangle}{dt} = \frac{(H_{21} - H_{12})}{\hbar}\langle \hat{S}_x \rangle + \frac{i(H_{21} + H_{12})}{\hbar}(-\langle \hat{S}_y \rangle)$$
$$- \frac{i}{T_1}(\langle \hat{S}_z \rangle^{eq} - \langle \hat{S}_z \rangle),$$

(5.7.38)

and with a reordering in the first term on the right,

$$\frac{d\langle \hat{S}_z \rangle}{dt} = \frac{(H_{12} - H_{21})}{i\hbar}\langle \hat{S}_x \rangle + \frac{(H_{21} + H_{12})}{\hbar}\langle \hat{S}_y \rangle + \frac{1}{T_1}(\langle \hat{S}_z \rangle^{eq} - \langle \hat{S}_z \rangle).$$

(5.7.39)

Since the Hamiltonian is Hermitian, $H_{12} = (H_{21})^*$ and $(H_{12} - H_{21})/i$ is real. Thus, Eqs. (5.7.35), (5.7.37) and (5.7.39) have only real quantities and form the Bloch Equations.

Yet another transformation is helpful. Yes, one more to make contact with a common form of the Bloch Equations! We define the vector

$$\langle \hat{\bar{S}} \rangle = \langle \hat{S}_x \rangle \hat{x} + \langle \hat{S}_y \rangle \hat{y} + \langle \hat{S}_z \rangle \hat{z}, \qquad (5.7.40)$$

in terms of the three orthogonal unit vectors. We next introduce a second vector

$$\bar{\omega} = \frac{(H_{12} + H_{21})}{\hbar} \hat{x} - \frac{(H_{12} - H_{21})}{i\hbar} \hat{y} - \frac{\Delta E}{\hbar} \hat{z}. \qquad (5.7.41)$$

This unusual choice of components in the vector $\bar{\omega}$ allows us to rewrite Eqs. (5.7.35), (5.7.37) and (5.7.39) in terms of the vector cross-product

$$\frac{d\langle \hat{S}_x \rangle}{dt} = (\bar{\omega} \times \hat{\bar{S}})_x - \frac{\langle \hat{S}_x \rangle}{T_2}, \qquad (5.7.42)$$

$$\frac{d\langle \hat{S}_y \rangle}{dt} = (\bar{\omega} \times \hat{\bar{S}})_y - \frac{\langle \hat{S}_y \rangle}{T_2}, \qquad (5.7.43)$$

$$\frac{d\langle \hat{S}_z \rangle}{dt} = (\bar{\omega} \times \hat{\bar{S}})_z + \frac{(\langle \hat{S}_z \rangle^{eq} - \langle \hat{S}_z \rangle)}{T_1}. \qquad (5.7.44)$$

These are the Bloch Equations as first set forth by Bloch (1946). In Sec. 5.8 we solve them for a simple case.

5.8. A Solution to the Bloch Equations

We consider a magnetic resonance example for a spin-$\frac{1}{2}$ particle with a constant magnetic field in the z-direction. We look when the time-dependent interaction has been turned off at time t_0, so we have free spin precession. This topic was introduced in Sec. 1.3 and in the Rambles of Chapter 1. Blum (1996, Sec. 8.4.2) discusses spin precession in terms of the Bloch Equations. Now we set

$$H_{12} = H_{21} = 0, \qquad (5.8.1)$$

so further state to state transitions are due to the relaxation terms in the Bloch Equations. We start with the Bloch Equations given in Eqs. (5.7.42) to (5.7.44) and note first that the vector $\bar{\omega}$ defined in

Eq. (5.7.41) now collapses to its $\hat{z}$ component,

$$\bar{\omega} = -\frac{\Delta E}{\hbar}\hat{z} = -\frac{(H_{22} - H_{11})}{\hbar}\hat{z}. \tag{5.8.2}$$

This simplifies the Bloch Equations which are

$$\frac{d\langle \hat{S}_x \rangle}{dt} = \frac{\Delta E}{\hbar}\langle \hat{S}_y \rangle - \frac{\langle \hat{S}_x \rangle}{T_2}, \tag{5.8.3}$$

$$\frac{d\langle \hat{S}_y \rangle}{dt} = -\frac{\Delta E}{\hbar}\langle \hat{S}_x \rangle - \frac{\langle \hat{S}_y \rangle}{T_2}, \tag{5.8.4}$$

$$\frac{d\langle \hat{S}_z \rangle}{dt} = \frac{1}{T_1}(\langle \hat{S}_z \rangle^{eq} - \langle \hat{S}_z \rangle). \tag{5.8.5}$$

The same equations result from Eqs. (5.7.35), (5.7.37) and (5.7.39).

We observe that the equation for $\langle \hat{S}_z \rangle$ is decoupled from the other two components and it has the same structure as Eq. (5.5.12). This allows us to write the solution for $t > t_0$

$$\langle \hat{S}_z(t) \rangle = \langle \hat{S}_z \rangle^{eq} + (\langle \hat{S}_z(t_0) \rangle - \langle \hat{S}_z \rangle^{eq})e^{-(t-t_0)/T_1}. \tag{5.8.6}$$

Thus, we see that the relaxation time T_1 does govern the approach of $\langle \hat{S}_z(t) \rangle$ to its equilibrium value.

We are left with the coupled Eqs. (5.8.3) and (5.8.4), which describe how $\langle \hat{S}_x(t) \rangle$ and $\langle \hat{S}_y(t) \rangle$ develop for times $t > t_0$. We define the angular frequency ω_0 by

$$\omega_0 = \Delta E/\hbar, \tag{5.8.7}$$

and for magnetic resonance ω_0 is the angular Larmor frequency for the constant magnetic field in the z-direction. Equations (5.8.3) and (5.8.4) are rewritten as

$$\frac{d\langle \hat{S}_x \rangle}{dt} = \omega_0\langle \hat{S}_y \rangle - \frac{\langle \hat{S}_x \rangle}{T_2}, \tag{5.8.8}$$

$$\frac{d\langle \hat{S}_y \rangle}{dt} = -\omega_0\langle \hat{S}_x \rangle - \frac{\langle \hat{S}_y \rangle}{T_2}. \tag{5.8.9}$$

Such coupled equations are encountered in Chaps. 2 to 4. We start with the time derivative of Eq. (5.8.8)

$$\frac{d^2\langle \hat{S}_x \rangle}{dt^2} = \omega_0\frac{d\langle \hat{S}_y \rangle}{dy} - \frac{1}{T_2}\frac{d\langle \hat{S}_x \rangle}{dt}, \tag{5.8.10}$$

and bring in Eq. (5.8.9) to find

$$\frac{d^2\langle \hat{S}_x \rangle}{dt^2} = \omega_0 \left(-\omega_0 \langle \hat{S}_x \rangle - \frac{\langle \hat{S}_y \rangle}{T_2} \right) - \frac{1}{T_2} \frac{d\langle \hat{S}_x \rangle}{dt}. \qquad (5.8.11)$$

We turn to Eq. (5.8.8) to substitute for $\langle \hat{S}_y \rangle$ and find

$$\frac{d^2\langle \hat{S}_x \rangle}{dt^2} = -\omega_0^2 \langle \hat{S}_x \rangle - \frac{1}{T_2} \left(\frac{d\langle \hat{S}_x \rangle}{dt} + \frac{\langle \hat{S}_x \rangle}{T_2} \right) - \frac{1}{T_2} \frac{d\langle \hat{S}_x \rangle}{dt}, \qquad (5.8.12)$$

which we rewrite in a standard form

$$\frac{d^2\langle \hat{S}_x \rangle}{dt^2} + \frac{2}{T_2} \frac{d\langle \hat{S}_x \rangle}{dt} + \left(\omega_0^2 + \frac{1}{T_2^2} \right) \langle \hat{S}_x \rangle = 0. \qquad (5.8.13)$$

We try a solution based on $e^{\beta t}$ and find the equation for β

$$\beta^2 + \frac{2}{T_2} \beta + \left(\omega_0^2 + \frac{1}{T_2^2} \right) = 0, \qquad (5.8.14)$$

and this quadratic equation yields

$$\beta = -\frac{1}{T_2} \pm i\omega_0. \qquad (5.8.15)$$

We find the same result for β, if we start with the time derivative of Eq. (5.8.9).

We see our solutions have both an exponential damping and an oscillation in time. We choose a sine and a cosine for the oscillatory terms, but we still need to match our initial conditions of $\langle \hat{S}_x(t_0) \rangle$ and $\langle \hat{S}_y(t_0) \rangle$. A possible choice is

$$\langle \hat{S}_x(t) \rangle = a \sin[\omega_0(t - t_0) + \phi] e^{-(t-t_0)/T_2}, \qquad (5.8.16)$$

$$\langle \hat{S}_y(t) \rangle = a \cos[\omega_0(t - t_0) + \phi] e^{-(t-t_0)/T_2}. \qquad (5.8.17)$$

The two parameters a and ϕ come from

$$\langle \hat{S}_x(t_0) \rangle = a \sin \phi, \qquad (5.8.18)$$

$$\langle \hat{S}_y(t_0) \rangle = a \cos \phi. \qquad (5.8.19)$$

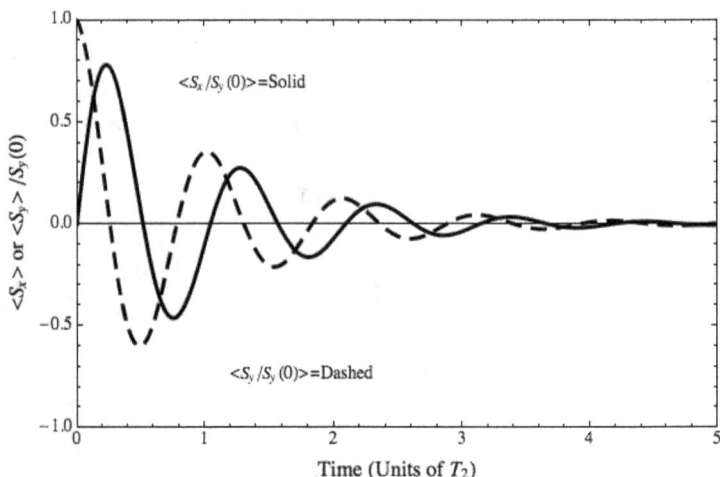

Figure 5.4. The normalized spin averages versus time. $\langle \hat{S}_x(t)/\hat{S}_y(0)\rangle$ is solid and $\langle \hat{S}_y(t)/\hat{S}_y(0)\rangle$ is dashed. We have $\phi = 0.0$, $t_0 = 0.0$ and $\omega_0 = 6.0/T_2$. The plot labels lack the $\char"5E$.

Equations (5.8.16) and (5.8.17) show how the projection of the spin angular momentum onto the x-y plane rotates at angular frequency ω_0 and is damped by the exponential with the relaxation time T_2.

    This behavior is illustrated in Fig. 5.4, where we plot $\langle \hat{S}_x(t)\rangle$ and $\langle \hat{S}_y(t)\rangle$. We set the phase $\phi = 0.0$, so $\langle \hat{S}_x(t_0)\rangle = 0.0$ and $\langle \hat{S}_y(t_0)\rangle = a$. We set $t_0 = 0.0$ for ease and express the time in units of T_2. Finally, we define $\omega_0 = 6.0/T_2$. Figure 5.4 shows $\langle \hat{S}_x(t)\rangle$ and $\langle \hat{S}_y(t)\rangle$ with both normalized to $\langle \hat{S}_y(t_0)\rangle$. We see both spin averages oscillate with time and are damped, due to the exponential relaxation terms that are dependent on T_2. Both spin averages approach 0.0 when the time increases.

    Thus, during free precession, the average values of the spin angular momentum components decay according to the relaxation times T_1 and T_2 within Eqs. (5.8.6), (5.8.16) and (5.8.17). We may convert the $\langle \hat{S}_i\rangle$ to the magnetization components $\langle \hat{M}_i\rangle$ by using the appropriate factors as in Eq. (1.3.11).

    We next recover the off-diagonal density matrix elements ρ_{12} and ρ_{21}. We go back through the chain of transformations starting with Eqs. (5.7.30) and (5.7.31). These are followed by Eqs. (5.7.8) and

(5.7.9) and we have

$$\langle \hat{S}_x(t) \rangle = \frac{\hbar}{2} U_1(t) = \frac{\hbar}{2} (\rho_{21}(t) + \rho_{12}(t)), \qquad (5.8.20)$$

$$\langle \hat{S}_y(t) \rangle = -\frac{\hbar}{2} U_2(t) = -i\frac{\hbar}{2} (\rho_{21}(t) - \rho_{12}(t)). \qquad (5.8.21)$$

We solve these two equations and find

$$\rho_{12}(t) = \frac{1}{\hbar} (\langle \hat{S}_x(t) \rangle - i\langle \hat{S}_y(t) \rangle), \qquad (5.8.22)$$

$$\rho_{21}(t) = \frac{1}{\hbar} (\langle \hat{S}_x(t) \rangle + i\langle \hat{S}_y(t) \rangle). \qquad (5.8.23)$$

And, yes, we could be more direct and use Eq. (5.7.26). In any event, we substitute Eqs. (5.8.16) and (5.8.17) into the last two equations. We learn for $t > t_0$ that

$$\rho_{12}(t) = \frac{a}{\hbar} \{\sin[\omega_0(t - t_0) + \phi] - i\cos[\omega_0(t - t_0) + \phi]\}e^{-(t-t_0)/T_2}, \qquad (5.8.24)$$

so

$$\rho_{12}(t) = \frac{a}{i\hbar} e^{i[\omega_0(t-t_0)+\phi]} e^{-(t-t_0)/T_2}. \qquad (5.8.25)$$

And

$$\rho_{21}(t) = \frac{a}{\hbar} \{\sin[\omega_0(t - t_0) + \phi] + i\cos[\omega_0(t - t_0) + \phi]\}e^{-(t-t_0)/T_2}, \qquad (5.8.26)$$

so

$$\rho_{21}(t) = \frac{ia}{\hbar} e^{-i[\omega_0(t-t_0)+\phi]} e^{-(t-t_0)/T_2}. \qquad (5.8.27)$$

We explicitly verify that

$$\rho_{12}(t) = (\rho_{21}(t))^*. \qquad (5.8.28)$$

Finally, Eqs. (5.7.32) and (5.7.10) provide

$$\langle \hat{S}_z(t) \rangle = -\frac{\hbar}{2} U_3(t) = -\frac{\hbar}{2} (\rho_{22}(t) - \rho_{11}(t)) = \frac{\hbar}{2} (2\rho_{11}(t) - 1). \qquad (5.8.29)$$

This connects the occupation probabilities of states $|1\rangle$ and $|2\rangle$ with the average value of $\langle \hat{S}_z \rangle$.

When $\rho_{11}(t) = 1$, $\langle \hat{S}_z(t) \rangle = \hbar/2$, while $\rho_{11}(t) = 0$ leads to $\langle \hat{S}_z(t) \rangle = -\hbar/2$.

5.9. More on the Density Matrix

This chapter introduces the density matrix and shows several applications. The density matrix is found to be useful in several fields of Physics. Blum (1996; 2012) shows how density matrices are used to describe atomic and molecular processes. Quantum Optics is home to many uses of the density matrix. Cohen-Tannoudji *et al.* (1992) and Rand (2010) are excellent sources. Earlier sections of this chapter make clear that magnetic resonance is often formulated through the density matrix and the Bloch Equations. Slichter (2010), or its earlier editions, is a good starting point for these subjects. In addition, the references cited in this chapter contain further instances of the density matrix. For example, Schatz and Ratner (2002, Chap. 11) discuss perturbative treatments; these approaches are based on the Liouville–von Neumann Equations in the Interaction Picture, the present Eq. (5.3.35), which is formally solved as an integral equation. The latter allows iteration and the solution of the Liouville–von Neumann Equations to the desired order.

Finally, we referred to master equations when relaxation was introduced in Sec. 5.5. This topic is discussed more fully and in useful detail by Blum (1996; 2012) and Shore (1990, Secs. 6.4 and 6.9).

The next section contains Ramble 5 and presents a direct solution for the time-dependent density matrix elements.

5.10. Ramble 5: A Solution of the Time-Dependent Equations for the Density Matrix Elements

Author's Note: The following derivation was kindly provided in July 2011 by Prof. Michael D. Fayer and his Research Group at Stanford University. I present the derivation in the notation of Sec. 5.4. I thank Prof. Fayer for this solution.

We start with the Liouville–von Neumann Equations for a two-level system in the Interaction Picture with the Rotating Wave Approximation (RWA). For convenience, we bring Eqs. (5.4.12) to (5.4.15) here

$$i\frac{d\rho_{I11}}{dt} = \frac{\tilde{\Omega}}{2}(e^{i(\omega-\omega_0)t}\rho_{I21} - e^{-i(\omega-\omega_0)t}\rho_{I12}), \qquad (5.10.1)$$

$$i\frac{d\rho_{I12}}{dt} = \frac{\tilde{\Omega}}{2}e^{i(\omega-\omega_0)t}(\rho_{I22} - \rho_{I11}), \qquad (5.10.2)$$

$$i\frac{d\rho_{I21}}{dt} = \frac{\tilde{\Omega}}{2}e^{-i(\omega-\omega_0)t}(\rho_{I11} - \rho_{I22}), \qquad (5.10.3)$$

$$i\frac{d\rho_{I22}}{dt} = \frac{\tilde{\Omega}}{2}(e^{-i(\omega-\omega_0)t}\rho_{I12} - e^{i(\omega-\omega_0)t}\rho_{I21}). \qquad (5.10.4)$$

We start the two-level system in the lower energy state

$$\rho_{I11}(t = 0) = 1, \qquad (5.10.5)$$

and

$$\rho_{I12}(t = 0) = \rho_{I21}(t = 0) = \rho_{I22}(t = 0) = 0. \qquad (5.10.6)$$

In addition, we repeat the definitions

$$\delta\omega = \omega - \omega_0, \qquad (5.10.7)$$

and

$$\Omega = [(\delta\omega)^2 + \tilde{\Omega}^2]^{1/2}. \qquad (5.10.8)$$

We begin with the incorporation of the phase factors. We set

$$\tilde{\rho}_{11} = \rho_{I11}, \qquad (5.10.9)$$

$$\tilde{\rho}_{22} = \rho_{I22}, \qquad (5.10.10)$$

$$\tilde{\rho}_{12} = e^{-i\delta\omega t}\rho_{I12}, \qquad (5.10.11)$$

and

$$\tilde{\rho}_{21} = e^{i\delta\omega t}\rho_{I21}. \qquad (5.10.12)$$

With these definitions, our differential equations become

$$\frac{d\tilde{\rho}_{11}}{dt} = \frac{i\tilde{\Omega}}{2}(\tilde{\rho}_{12} - \tilde{\rho}_{21}),$$
(5.10.13)

$$\frac{d\tilde{\rho}_{12}}{dt} = \frac{i\tilde{\Omega}}{2}(\tilde{\rho}_{11} - \tilde{\rho}_{22}) - i\delta\omega\tilde{\rho}_{12},$$
(5.10.14)

$$\frac{d\tilde{\rho}_{21}}{dt} = -\frac{i\tilde{\Omega}}{2}(\tilde{\rho}_{11} - \tilde{\rho}_{22}) + i\delta\omega\tilde{\rho}_{21},$$
(5.10.15)

$$\frac{d\tilde{\rho}_{22}}{dt} = -\frac{i\tilde{\Omega}}{2}(\tilde{\rho}_{12} - \tilde{\rho}_{21}).$$
(5.10.16)

We assume the solutions are

$$\tilde{\rho}_{ij}(t) = \tilde{\rho}_{ij}(0)e^{\lambda t},$$
(5.10.17)

and we need to solve for λ. Please note that the $\tilde{\rho}_{ij}(0)$ are constant coefficients. We substitute our guess into Eqs. (5.10.13) to (5.10.16) and find

$$\lambda\tilde{\rho}_{11}(0) = \frac{i\tilde{\Omega}}{2}(\tilde{\rho}_{12}(0) - \tilde{\rho}_{21}(0)),$$
(5.10.18)

$$\lambda\tilde{\rho}_{12}(0) = \frac{i\tilde{\Omega}}{2}(\tilde{\rho}_{11}(0) - \tilde{\rho}_{22}(0)) - i\delta\omega\tilde{\rho}_{12}(0),$$
(5.10.19)

$$\lambda\tilde{\rho}_{21}(0) = -\frac{i\tilde{\Omega}}{2}(\tilde{\rho}_{11}(0) - \tilde{\rho}_{22}(0)) + i\delta\omega\tilde{\rho}_{21}(0),$$
(5.10.20)

$$\lambda\tilde{\rho}_{22}(0) = -\frac{i\tilde{\Omega}}{2}(\tilde{\rho}_{12}(0) - \tilde{\rho}_{21}(0)),$$
(5.10.21)

and we have canceled the common exponential factor. We recast the last 4 equations as a set of homogenous linear equations for the $\tilde{\rho}_{ij}(0)$,

$$\begin{pmatrix} \lambda & -i\tilde{\Omega}/2 & i\tilde{\Omega}/2 & 0 \\ -i\tilde{\Omega}/2 & \lambda+i\delta\omega & 0 & i\tilde{\Omega}/2 \\ i\tilde{\Omega}/2 & 0 & \lambda-i\delta\omega & -i\tilde{\Omega}/2 \\ 0 & i\tilde{\Omega}/2 & -i\tilde{\Omega}/2 & \lambda \end{pmatrix} \begin{pmatrix} \tilde{\rho}_{11}(0) \\ \tilde{\rho}_{12}(0) \\ \tilde{\rho}_{21}(0) \\ \tilde{\rho}_{22}(0) \end{pmatrix} = \begin{pmatrix} 0 \\ 0 \\ 0 \\ 0 \end{pmatrix}.$$
(5.10.22)

The determinant equals zero when

$$\lambda^2(\lambda^2 + (\delta\omega)^2 + \tilde{\Omega}^2) = 0, \tag{5.10.23}$$

which yields

$$\lambda = 0, \tag{5.10.24}$$

and

$$\lambda = \pm i\sqrt{(\delta\omega)^2 + \tilde{\Omega}^2} = \pm i\Omega. \tag{5.10.25}$$

We first solve for $\tilde{\rho}_{11}(t)$, which we write as

$$\tilde{\rho}_{11}(t) = A_{11} + B_{11}e^{i\Omega t} + C_{11}e^{-i\Omega t}. \tag{5.10.26}$$

Equation (5.10.26) at $t = 0$ provides one equation for the three unknowns and leads to two more equations for the first two derivatives at $t = 0$. These are evaluated through Eqs. (5.10.13) to (5.10.15) with the help of the initial conditions given by Eqs. (5.10.5) and (5.10.6).

The first condition is

$$\tilde{\rho}_{11}(0) = 1 = A_{11} + B_{11} + C_{11}. \tag{5.10.27}$$

The first derivative yields

$$\frac{d\tilde{\rho}_{11}(t = 0)}{dt} = 0 = i\Omega B_{11} - i\Omega C_{11}. \tag{5.10.28}$$

or

$$B_{11} = C_{11}. \tag{5.10.29}$$

The second derivative provides

$$\frac{d^2\tilde{\rho}_{11}(t = 0)}{dt^2} = -\Omega^2(B_{11} + C_{11}) = \frac{i\tilde{\Omega}}{2}\left(\frac{d\tilde{\rho}_{12}(0)}{dt} - \frac{d\tilde{\rho}_{21}(0)}{dt}\right), \tag{5.10.30}$$

and the last equality gives $(i\tilde{\Omega}/2)(2i\tilde{\Omega}/2) = -\tilde{\Omega}^2/2$. So, the third relation is

$$\Omega^2(B_{11} + C_{11}) = 2\Omega^2 B_{11} = \tilde{\Omega}^2/2. \qquad (5.10.31)$$

Thus,

$$B_{11} = C_{11} = \frac{\tilde{\Omega}^2}{4\Omega^2}, \qquad (5.10.32)$$

and with Eq. (5.10.27), we have

$$A_{11} = 1 - 2B_{11} = 1 - \frac{\tilde{\Omega}^2}{2\Omega^2}. \qquad (5.10.33)$$

Finally, we write Eq. (5.10.26) as

$$\tilde{\rho}_{11}(t) = 1 - \frac{\tilde{\Omega}^2}{2\Omega^2} + \frac{\tilde{\Omega}^2}{4\Omega^2}(e^{i\Omega t} + e^{-i\Omega t}) = 1 + \frac{\tilde{\Omega}^2}{2\Omega^2}(\cos \Omega t - 1),$$
$$(5.10.34)$$

which we convert to

$$\tilde{\rho}_{11}(t) = \rho_{I11}(t) = 1 - \frac{\tilde{\Omega}^2}{\Omega^2} \sin^2(\Omega t/2). \qquad (5.10.35)$$

We immediately have

$$\tilde{\rho}_{22}(t) = \rho_{I22}(t) = 1 - \rho_{I11}(t) = \frac{\tilde{\Omega}^2}{\Omega^2} \sin^2(\Omega t/2). \qquad (5.10.36)$$

It remains to find the off-diagonal terms. We determine $\tilde{\rho}_{12}(t)$ by a similar approach. The initial value yields

$$\tilde{\rho}_{12}(0) = 0 = A_{12} + B_{12} + C_{12}, \qquad (5.10.37)$$

and the first derivative produces

$$\frac{d\tilde{\rho}_{12}(t = 0)}{dt} = \frac{i\tilde{\Omega}}{2} = i\Omega(B_{12} - C_{12}), \qquad (5.10.38)$$

with the aid of the initial conditions given in Eqs. (5.10.5) and (5.10.6). The second derivative provides

$$\frac{d^2\tilde{\rho}_{12}(t=0)}{dt^2} = -\Omega^2(B_{12} + C_{12}) = -i\delta\omega\frac{i\tilde{\Omega}}{2} = \delta\omega\frac{\tilde{\Omega}}{2}. \quad (5.10.39)$$

Equations (5.10.38) and (5.10.39) become

$$B_{12} - C_{12} = \frac{\tilde{\Omega}}{2\Omega}, \quad (5.10.40)$$

and

$$B_{12} + C_{12} = -\delta\omega\frac{\tilde{\Omega}}{2\Omega^2}. \quad (5.10.41)$$

These are solved to find

$$B_{12} = \frac{\tilde{\Omega}}{4\Omega}\left(1 - \frac{\delta\omega}{\Omega}\right), \quad (5.10.42)$$

$$C_{12} = -\frac{\tilde{\Omega}}{4\Omega}\left(1 + \frac{\delta\omega}{\Omega}\right), \quad (5.10.43)$$

and

$$A_{12} = \delta\omega\frac{\tilde{\Omega}}{2\Omega^2}. \quad (5.10.44)$$

We now put Eqs. (5.10.42) to (5.10.44) together

$$\tilde{\rho}_{12}(t) = \frac{\delta\omega\tilde{\Omega}}{2\Omega^2} + \frac{\tilde{\Omega}}{4\Omega}\left(1 - \frac{\delta\omega}{\Omega}\right)e^{i\Omega t} - \frac{\tilde{\Omega}}{4\Omega}\left(1 + \frac{\delta\omega}{\Omega}\right)e^{-i\Omega t},$$

$$(5.10.45)$$

and regroup the exponentials, so

$$\tilde{\rho}_{12}(t) = \frac{\delta\omega\tilde{\Omega}}{2\Omega^2} + \frac{\tilde{\Omega}}{4\Omega}(e^{i\Omega t} - e^{-i\Omega t}) - \frac{\delta\omega\tilde{\Omega}}{4\Omega^2}(e^{i\Omega t} + e^{-i\Omega t}).$$

$$(5.10.46)$$

In turn, this becomes

$$\tilde{\rho}_{12}(t) = \frac{i\tilde{\Omega}}{2\Omega}(\sin\Omega t) - \frac{\delta\omega\tilde{\Omega}}{2\Omega^2}(\cos\Omega t - 1)$$

$$= \frac{i\tilde{\Omega}}{2\Omega}(\sin\Omega t) + \frac{\delta\omega\tilde{\Omega}}{\Omega^2}(\sin^2(\Omega t/2)). \quad (5.10.47)$$

Finally, with Eq. (5.10.11), we find

$$\rho_{I12}(t) = e^{i\delta\omega t}\tilde{\rho}_{12}(t) = e^{i\delta\omega t}\left(\frac{i\tilde{\Omega}}{2\Omega}(\sin\Omega t) + \frac{\delta\omega\tilde{\Omega}}{\Omega^2}(\sin^2(\Omega t/2))\right),$$

$$(5.10.48)$$

and by Eq. (5.3.16)

$$\rho_{I21}(t) = e^{-i\delta\omega t}\left(-\frac{i\tilde{\Omega}}{2\Omega}(\sin\Omega t) + \frac{\delta\omega\tilde{\Omega}}{\Omega^2}(\sin^2(\Omega t/2))\right). \quad (5.10.49)$$

Thus, we have worked through the solution provided by Prof. Michael D. Fayer and we now see how Eqs. (5.4.20) to (5.4.23) arise.

We next enter Chap. 6 and the world of resonance fluorescence of two-level systems.

References

E. Beham, A. Zrenner, F. Findeis, M. Bichler, and G. Abstreiter, "Rabi-flopping of the ground state exciton in a single self-assembled quantum dot", *Phys. Stat. Sol. B* **238**, 366–369 (2003).

E. R. Bittner, *Quantum Dynamics: Applications in Biological and Materials Systems* (CRC Press, Boca Raton, 2010).

F. Bloch, "Nuclear Induction", *Phys. Rev.* **70**, 460–474 (1946)

K. Blum, *Density Matrix Theory and Applications*, 2nd ed. (Plenum Press, New York, 1996).

K. Blum, *Density Matrix Theory and Applications*, 3rd ed. (Springer-Verlag, New York, 2012).

H. S. Brandi, A. Latgé, Z. Barticevic, and L. E. Oliveira, "Rabi oscillations in two-level semi-conductor systems", *Solid State Communications* **135**, 386–389 (2005).

B. H. Bransden and C. J. Joachain, *Quantum Mechanics*, 2nd ed. (Prentice Hall, Harlow, England, 2000).

T. Chakraborty, *Quantum Dots: A Survey of the Properties of Artificial Atoms* (Elsevier Science, Amsterdam, 1999).

C. Cohen-Tannoudji, J. Dupont-Roc, and G. Grynberg, *Atom-Photon Interactions: Basic Processes and Applications* (John Wiley and Sons, New York, 1992).

D. Dubbers and H.-J. Stöckmann, *Quantum Physics: The Bottom-Up Approach* (Springer-Verlag, Berlin, 2013).

M. D. Fayer, *Elements of Quantum Mechanics* (Oxford University Press, New York, 2001).

G. W. Hanson, *Fundamentals of Nanoelectronics* (Pearson Prentice Hall, Upper Saddle River, N. J., 2008).

S. Nadj-Perge, S. M. Frolov, E. P. A. M. Bakkers, and L. P. Kouwenhoven, "Spin-orbit qubit in a semiconductor nanowire", *Nature* **468**, 1084–1087 (2010).

S. C. Rand, *Lectures on Light: Nonlinear and Quantum Optics using the Density Matrix* (Oxford University Press, Oxford, 2010).

J. J. Sakurai and Jim Napolitano, *Modern Quantum Mechanics*, 2nd ed. (Addison-Wesley, Boston, 2011).

G. C. Schatz and M. A. Ratner, *Quantum Mechanics in Chemistry* (Dover Publications, Mineola, New York, 2002).

B. W. Shore, *The Theory of Coherent Atomic Excitation*, Vol. 1 Simple Atoms and Fields (John Wiley and Sons, New York, 1990).

R. Skomski, *Simple Models of Magnetism* (Oxford University Press, Oxford, 2008), Paperback 2012.

C. P. Slichter, *Principles of Magnetic Resonance*, 3rd ed. (Springer-Verlag, New York, 2010).

D. W. Snoke, *Solid State Physics: Essential Concepts* (Addison-Wesley, San Francisco, 2009).

S. Stenholm, *Foundations of Laser Spectroscopy* (Dover Publications, Mineola, New York, 2005).

A. Zrenner, E. Beham, S. Stufler, F. Findeis, M. Bichler, and G. Abstreiter, "Coherent properties of a two-level system based on a quantum-dot photo-diode", *Nature* **418**, 612–614 (2002).

Chapter 6

The Second-Order Correlation
Function for Two-Level Systems

We return to optical excitation and optical emission, and we explore experiments that focus on the photons emitted when a two-level system returns to its ground state. Such experiments are examples of resonance fluorescence and we find the density matrices of Chap. 5 useful in untangling the experiments. Our discussions involve systems such as defects in diamonds, quantum dots and molecules in condensed matter. Much of this work is aimed at developing single-photon sources, which would be useful for quantum computing and integrated circuits. For the latter, photons replace electrons for some of the "communication" levels on-chip and between chips. This requires controllable excitation of the system and the emission of a single photon upon demand. Aharonovich *et al.* (2011) review single-photon emission by diamond defects, while Migdal *et al.* (2013) and Predojević and Mitchell (2015) provide additional information on a variety of systems. Kuhlmann *et al.* (2015) and Ding *et al.* (2016) are examples of the ongoing research into single-photon sources.

We explore the optical excitation of few-level systems in condensed matter. Such systems offer many routes for the relaxation of an excited state, with non-radiative paths being very common. The latter usually involve the excitation of phonons, an additional level or a nearby separate entity, which may be another defect or a separate device structure. In any case, our two-level system idealization is challenged and we need to include relaxation in our equations for the elements of the density matrix. In addition, we need to determine what measurements help us differentiate two-level systems from, say, three-level systems.

It is useful to consider the expected time dependences involved in experimental data or in proposed experiments. The sample times and intervals need to be fine enough to catch the expected physical mechanisms in action. This is especially so when the data have an on/off character,and a key question is the nature of the off-state. It is usually the behavior of an individual electron or spin that drives the transitions between on and off, and then between off and on. Two types of data are common. The first is optical emission data that resemble an on-off signal. Let us recall the random telegraph noise (RTN) signals discussed in Chaps. 2 and 4. The intrinsic lifetime of the excited state needs to be compared with the experiment's time resolution. If the intrinsic lifetime is much shorter than the time bin used in accumulating the optical emission data, then each on-time is composed of many emission events. The second type of data collects the time between two photon emissions, so finer time resolution is required. Resonance fluorescence provides an example of this type of experiment. Such data lead to second-order correlation functions.

Section 6.1 describes experiments based on emitted photons in which the measured data have an on and off character. Section 6.2 has further examples of emission data and focuses on whether the system is acting like a two-level or a three-level object. We introduce the second-order correlation function, $g^{(2)}(\tau)$, to help us decide. Here, τ is the time that has elapsed between the observation of the first emitted photon and the observation of a second photon. Section 6.3 develops $g^{(2)}(\tau)$ for a two-level system and presents numerical calculations based on the Liouville–von Neumann Equations introduced in Chap. 5. We explicitly include relaxation and compare the results with experimental data from Flagg *et al.* (2009). Then, in Sec. 6.4 we derive a common weak-field expression for $g^{(2)}(\tau)$ from our Sec. 6.3 result. We end with Sec. 6.5, which describes several challenging few-level problems to explore.

6.1. Two Levels versus Three Levels

We start with the data from a self-assembled InP quantum dot of Pistol *et al.* (1999). They made the quantum dots on a layer of GaInP on a GaAs substrate and the InP quantum dots have a GaInP capping layer. Pistol and colleagues explain that the following data are typical of only a small fraction of the dots they measured. Generally, an electron-hole pair is created when the quantum dot absorbs a photon. Then, if the electron and hole eventually recombine, a photon may be emitted. The photon energy is 1.66 eV for these InP quantum dots. Figure 6.1 is the emission intensity of a self-assembled InP quantum dot at 7 K (Pistol *et al.*, 1999, Fig. 2(a)). The data clearly show the appearance of random telegraph noise (RTN). Pistol and colleagues plotted the frequency of on- and off-times and found that their data are fit by a single exponential for each, as seen in Fig. 6.2. The temptation is to declare this a two-level system, due to the single exponential distributions. However, this raises the question of what the off-state is, since the on-times and off-times are fractions of a second to seconds. RTN data occur for quantum jumps observed in three-level systems by, for example, Bergquist *et al.* (1986), Nagourney *et al.* (1986) and Sauter *et al.* (1986). Here the third level is sometimes called the shelving state and provides the off-state. That is, the electron enters the shelving state and the optical emission is interrupted. This leads us to postulate that the photo-excited electron or hole in the InP quantum dot is trapped within the quantum dot, or has perhaps escaped for a period of time. Do we have two levels or three levels?

Basché *et al.* (1997, Fig. 12) present RTN data for a single terrylene molecule within p-terphenyl (see Fig. 6.3). Their off-time data show two exponentials with time constants of 0.43 and 3 ms. They note these results are similar to triplet state lifetimes for terrylene. Basché and colleagues state the on-time data are not sufficient to fit more than a single exponential, which has a time constant of 6.3 ms. In addition, they find the single exponential is not a good fit at long times. A return to quantum dots reveals a wide range of on-time and off-time distributions for similar

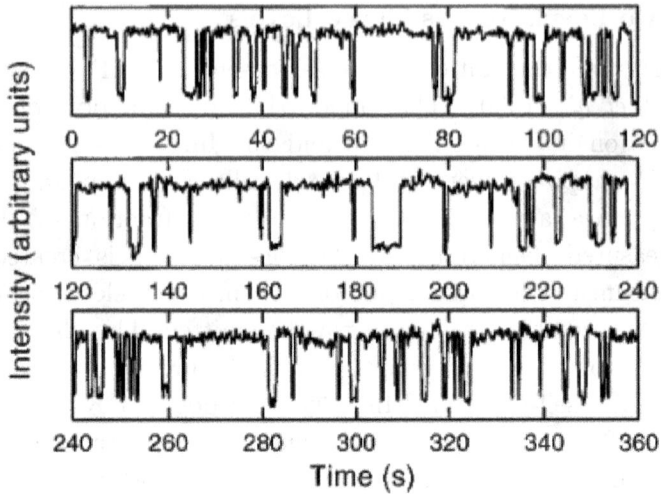

Figure 6.1. The emission intensity of a InP quantum dot. Random telegraph noise between two energy levels is observed (Pistol *et al.*, 1999, Fig. 2(a)). Reprinted figure with permission. Copyright 1999 by the American Physical Society.

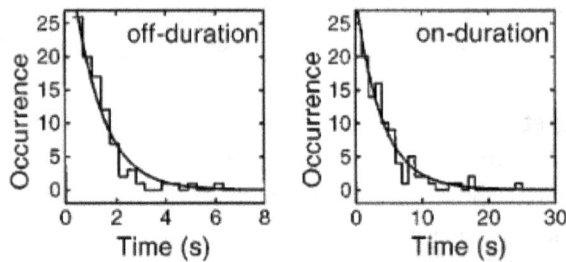

Figure 6.2. Histograms of the off-times (*left*) and the on-times (*right*) for a InP quantum dot (Pistol *et al.*, 1999, Figs. 2(b) and 2(c)). Reprinted figure with permission. Copyright 1999 by the American Physical Society.

dots. Kuno *et al.* (2001) deduce power law distributions for CdSe quantum dots, in contrast to the exponential distributions of Fig. 6.2. Figure 6.4 shows their measured on- and off-times (Kuno *et al.* 2001, Figs. 10(a) and 10(b)). Galland *et al.* (2011) find power laws and more complicated distributions for their measured on- and off-time distribution for CdSe/CdS quantum dots.

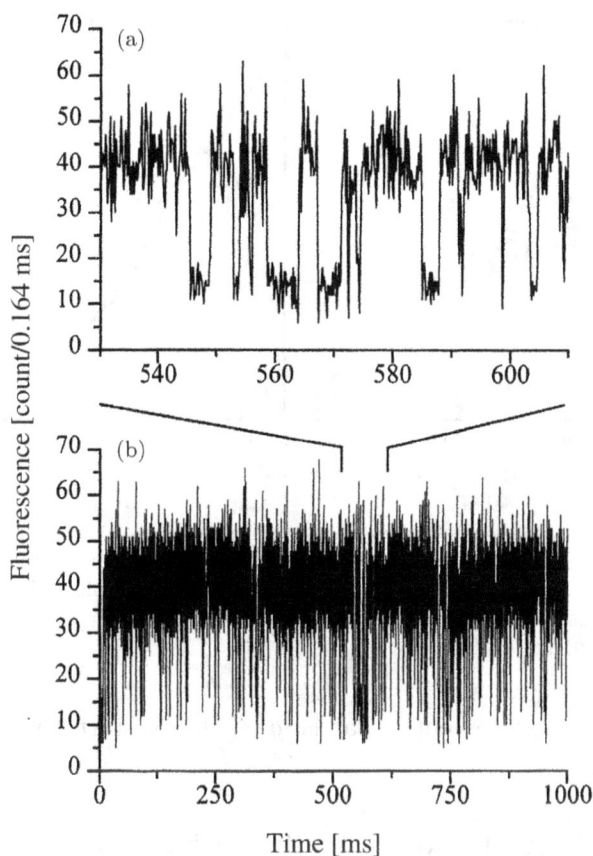

Figure 6.3. Quantum jump data for a single terrylene molecule in p-terphenyl as a function of time at T = 1.4 K. Part (a) shows 80 ms in detail (Basché *et al.*, 1997, Fig. 12). This figure is reproduced from *Single-Molecule Optical Detection, Imaging and Spectroscopy*, p. 52. Copyright Wiley-VCH Verlag GmbH & Co. KGaA. Reproduced with permission.

These few examples make clear that the analysis of RTN signals in condensed matter does not lead to simple interpretations. Kuno *et al.* (2001) invoke tunneling to help explain their measured power laws, while Stefani *et al.* (2009) suggest further charge separation reasons for power laws. Clearly, other measurements are needed to help us decide when a two-level system is present and when more levels are involved. A favorite approach is two-photon

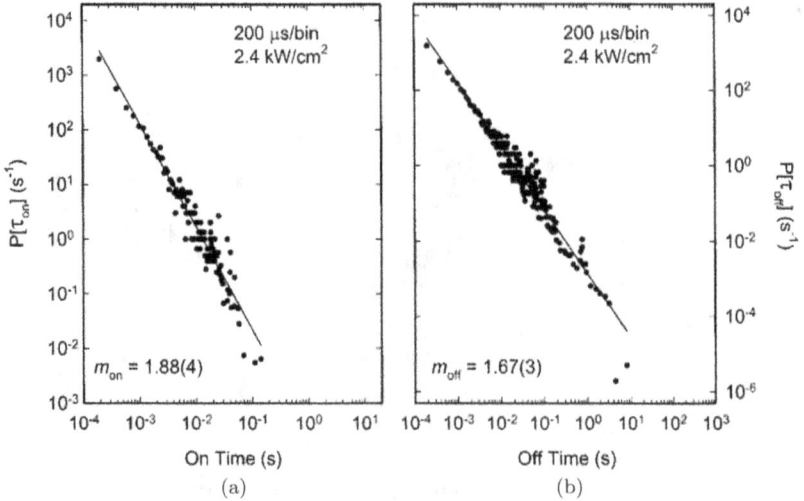

Figure 6.4. Distributions of on-times (a) and off-times (b) (Kuno *et al.*, 2001, Fig. 10). Power law distributions are seen on these log-log plots. The integration time is $200\,\mu s$/bin. Reprinted figure with permission. Copyright 2001 by the American Institute of Physics.

correlation measurements, that is, $g^{(2)}(\tau)$. We start down this path in Sec. 6.2.

6.2. The Second-Order Correlation Function and Resonance Fluorescence

Osad'ko (2003) and Basché *et al.* (1997) provide extensive details on the types of two-photon correlation experiments. A continuous wave laser is usually used to excite the system under study, and a means for photon detection is required. The time of the photon emission is recorded, and successive photon emissions are noted. After the experiment is concluded, the time intervals between successive photon emissions are binned and a histogram is constructed. Then any background signal is eliminated and the resulting counts are normalized by the long-time count rates. Detector efficiencies are worked in and usually instrument resolution needs to be taken into account. This becomes the basis for the second-order correlation function $g^{(2)}(\tau)$ with τ being the time interval or delay time. Both of the above references explain how calculations are connected with the experiments. In addition, texts on Quantum Optics (Loudon, 2000; Agarwal, 2013) are excellent resources for the simulation of $g^{(2)}(\tau)$. Further references are provided in the following sections.

Thus, the second-order correlation function, also called the second-order coherence, $g^{(2)}(\tau)$, involves the measurement of a photon at time zero and a second photon at time $t = \tau$. These emitted photons correspond to the transition from the excited state to the ground state of our two-level system. With the emission of the first photon at time zero, the system returns to its ground state. Then we need the system to absorb a photon and return to its excited state. Hence, we expect $g^{(2)}(\tau)$ to be proportional to the occupation probability of the excited state at $t = \tau$.

We begin with a two-level defect in diamond that emits at 734 nm, as described by Simpson *et al.* (2009). Figure 6.5 shows their data for $g^{(2)}(\tau)$ for this room-temperature emitter. The data are normalized to 1.0 at long times, as is standard for such plots. The data rise from a minimum at $\tau = 0$ and barely exceed 1.0, even when the incident pump power is increased by a factor of 10. The shape of these plots is explored in Sec. 6.3. Similar data for other defects in diamond appear in Fig. 6.6 (Aharonovich *et al.*, 2011, Figs. 7(c) and 7(d)). The left-hand plots are for a two-level defect that emits at 756 nm and the trends of Fig. 6.5 are seen. The successive plots are offset. The right-hand plots go with a defect that emits at 749 nm. Increased excitation power causes a noticeable peak to develop as the absolute value of τ increases. This peak turns out to be the signature of a system with more than 2 levels. The excited electron, or hole, makes a transition to the third energy level, which is the shelf state, to use the quantum jump terminology. The emission restarts when the defect returns to its ground state and a photon is then absorbed (Bergquist *et al.*, 1986). This sets the stage for the emission of the second photon. Thus, the location in time of the peak provides information on the lifetime of the shelf state.

Additional data appear in Sec. 6.3 where we develop expressions for $g^{(2)}(\tau)$ for a two-level system.

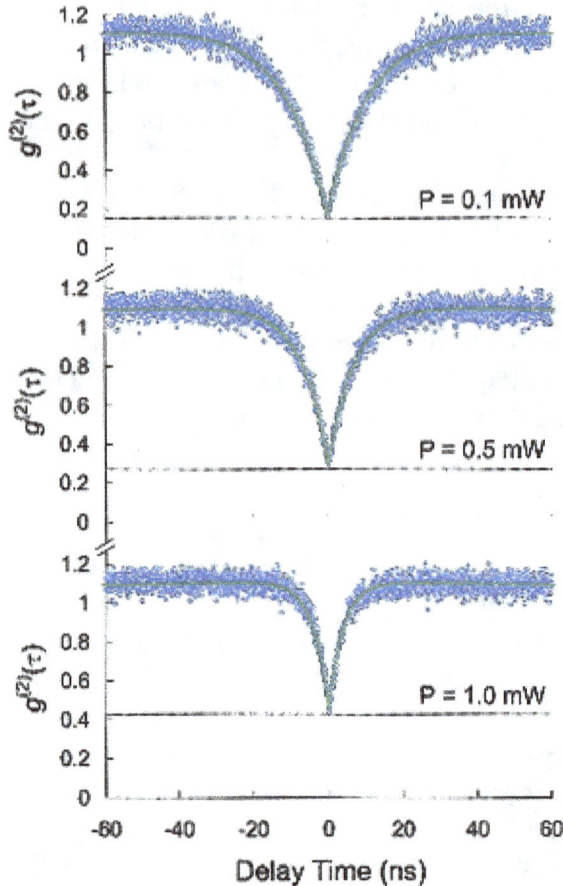

Figure 6.5. Two-photon correlation measurements for a two-level defect in diamond at room temperature for three incident pump powers (Simpson *et al.*, 2009, Fig. 2). Reprinted figure with permission. Copyright 2009 by the American Institute of Physics.

6.3. The Second-Order Correlation Function for a Two-Level System

For the resonance fluorescence measurements, we need the system to absorb a photon and return to its excited state. Hence, it is plausible that $g^{(2)}(\tau)$ is proportional to the occupation probability of the excited state at $t = \tau$. We work with the density matrix introduced

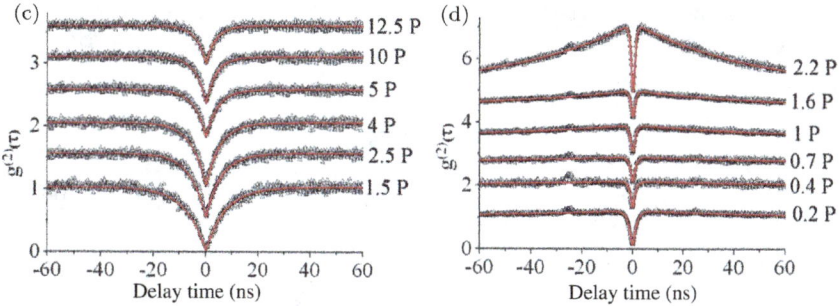

Figure 6.6. Two-photon correlation measurements for a two-level defect in diamond (c) and a three-level defect (d) in diamond at room temperature (Aharonovich *et al.*, 2011, Figs. 7(c) and 7(d)). Each curve is labeled with its excitation power with P the saturation excitation power. ©IOP Publishing. Reproduced with permission.

in Chap. 5 and solve the Liouville–von Neumann Equations. Our equations resemble those in Sec. 5.4, but we now label the ground state 0 and the excited state 1. This brings our notation into line with that of Flagg (2008). Its App. 1 provides a useful derivation of $g^{(2)}(\tau)$ through Laplace Transforms, which is an alternate approach to the one used in this section. Additional helpful material is found in Loudon (2000, Chaps. 2, 4, 7 and 8).

We start with an expression for $g^{(2)}(\tau)$ in terms of $\rho_{11}(\tau)$, which is the occupation probability of the excited state in the present notation,

$$g^{(2)}(\tau) = \rho_{11}(\tau)/\rho_{11}(\infty). \qquad (6.3.1)$$

This makes the long-time limit of $g^{(2)}(\tau)$ equal to 1. This equation is derived in App. 6 through an operator approach based on Loudon (2000, Secs. 4.9 and 7.9). We formulate the Liouville–von Neumann Equations and solve them directly with the aid of the Mathematica (2012) computer software. The present section concludes with computational results that parallel the results found for resonance fluorescence experiments on a quantum dot by Flagg (2008) and Flagg *et al.* (2009).

The Liouville–von Neumann Equations give us the elements of the density matrix. We assume resonance and have generalized

Eqs. (5.4.12) to (5.4.15) to include relaxation. We make the Rotating Wave Approximation (RWA) and use the Interaction Picture, although we drop the subscript I from the density matrix element subscripts. All of this leads to (Flagg, 2008)

$$\frac{d\rho_{00}}{dt} = \frac{i\Omega}{2}\rho_{01} - \frac{i\Omega}{2}\rho_{10} + \frac{1}{T_1}\rho_{11}, \qquad (6.3.2)$$

$$\frac{d\rho_{01}}{dt} = \frac{i\Omega}{2}\rho_{00} - \frac{1}{T_2}\rho_{01} - \frac{i\Omega}{2}\rho_{11}, \qquad (6.3.3)$$

$$\frac{d\rho_{10}}{dt} = -\frac{i\Omega}{2}\rho_{00} - \frac{1}{T_2}\rho_{10} + \frac{i\Omega}{2}\rho_{11}, \qquad (6.3.4)$$

$$\frac{d\rho_{11}}{dt} = -\frac{i\Omega}{2}\rho_{01} + \frac{i\Omega}{2}\rho_{10} - \frac{1}{T_1}\rho_{11}. \qquad (6.3.5)$$

Again we note the ground state is labeled 0 and the excited state is labeled 1.

Several remarks are in order. Firstly, before relaxation is introduced, ρ_{01} and ρ_{10} in these equations differ by a minus sign from the equations in Chap. 5. Secondly, the above density matrix elements here do not relax to equilibrium values. Instead, they approach steady–state values at long times, since we assume continuous laser illumination. Thirdly, T_1 is the lifetime of the excited state of the two-level system in the quantum dot. Fourthly, T_2 represents a dephasing or decoherence time for the quantum dot. This represents a deexcitation of the excited state to the ground state by paths other than optical emission. Finally, since we assume resonance, Ω is proportional to the strength of the illuminating laser light electric field and is the angular Rabi frequency. This leads to a Rabi energy of $\hbar\Omega$.

Before we solve Eqs. (6.3.2) to (6.3.5), we find the steady–state result for $\rho_{11}(\infty)$. This is needed for $g^{(2)}(\tau)$ and arises from setting the time derivatives to zero. Then Eq. (6.3.3) minus Eq. (6.3.4) yields

$$\rho_{01}(\infty) - \rho_{10}(\infty) = i\Omega T_2(\rho_{00}(\infty) - \rho_{11}(\infty)). \qquad (6.3.6)$$

We place this into Eq. (6.3.2) and use

$$\rho_{00}(t) + \rho_{11}(t) = 1, \qquad (6.3.7)$$

to find

$$\rho_{11}(\infty) = \frac{1}{2} \frac{\Omega^2}{\Omega^2 + (1/T_1 T_2)}. \tag{6.3.8}$$

This result also falls out of the time-dependent solution.

Now for $\rho_{11}(t)$, the occupation probability of the excited state. We first rewrite Eqs. (6.3.2) to (6.3.5) as a vector equation with the matrix M. This has the time-independent matrix elements from the right-hand sides of the equations and the vector $\bar{\rho}$ with the density matrix elements,

$$\frac{d\bar{\rho}(t)}{dt} = M\bar{\rho}(t). \tag{6.3.9}$$

We assume

$$\bar{\rho}(t) = \bar{I} e^{\lambda t}, \tag{6.3.10}$$

with the identity vector $\bar{I}$. This leads to the eigenvalue equation

$$\det |M - \lambda I| = 0, \tag{6.3.11}$$

with I the 4×4 matrix with ones on the diagonal. Equation (6.3.11) yields a fourth-order polynomial equation in λ and the routine Solve of Mathematica provides the roots, which are the eigenvalues. We define the following combinations of variables to simplify the results,

$$\eta = \frac{1}{2} \left(\frac{1}{T_1} + \frac{1}{T_2} \right), \tag{6.3.12}$$

and

$$\mu = \left(\Omega^2 - \frac{1}{4} \left(\frac{1}{T_1} - \frac{1}{T_2} \right)^2 \right)^{1/2}. \tag{6.3.13}$$

These parameters appear in Flagg *et al.* (2009) and Flagg (2008), although typographical errors are frequent in their equations involving η and μ.

With η and μ, the four eigenvalues are

$$\lambda_1 = 0, \quad \lambda_2 = -1/T_2, \quad \lambda_3 = -\eta - i\mu, \quad \text{and} \quad \lambda_4 = -\eta + i\mu. \tag{6.3.14}$$

Mathematica verifies that $\det M = 0$ for each of the eigenvalues. We now write $\rho_{11}(t)$ in terms of four terms with these eigenvalues and four unknown constants

$$\rho_{11}(t) = a_1 + a_2 e^{-t/T_2} + a_3 e^{-(\eta+i\mu)t} + a_4 e^{-(\eta-i\mu)t}. \qquad (6.3.15)$$

We next need four linear equations in terms of the four a_i. We base these equations upon the initial conditions, which we assume are

$$\rho_{00}(t = 0) = 1, \qquad (6.3.16)$$

$$\rho_{01}(t = 0) = \rho_{10}(t = 0) = \rho_{11}(t = 0) = 0. \qquad (6.3.17)$$

Thus, we start with the two-level system in the ground state.

The first equation comes from $\rho_{11}(0) = 0$ and is

$$a_1 + a_2 + a_3 + a_4 = 0. \qquad (6.3.18)$$

The next three equations come from evaluating the first, second and third time derivatives of $\rho_{11}(t)$ at $t = 0$. This is possible through the use of the initial conditions and Eqs. (6.3.2) to (6.3.5) with Eq. (6.3.15). The first time derivative, which is Eq. (6.3.5), yields the second equation

$$\frac{-a_2}{T_2} - a_3(\eta + i\mu) + a_4(-\eta + i\mu) = 0. \qquad (6.3.19)$$

The third equation comes from the second time derivative of $\rho_{11}(t)$ at $t = 0$ and is

$$\frac{a_2}{T_2^2} + (\eta + i\mu)^2 a_3 + (-\eta + i\mu)^2 a_4 = \Omega^2/2. \qquad (6.3.20)$$

The third time derivative provides the fourth equation

$$-\frac{a_2}{T_2^3} - (\eta + i\mu)^3 a_3 + (-\eta + i\mu)^3 a_4 = -\frac{\Omega^2}{2}\left(\frac{1}{T_1} + \frac{1}{T_2}\right). \qquad (6.3.21)$$

This set of four linear equations in four unknowns is solved by the Mathematica routine LinearSolve, with the results expressed with the assistance of Eq. (6.3.8),

$$a_1 = \rho_{11}(\infty), \tag{6.3.22}$$

$$a_2 = 0, \tag{6.3.23}$$

$$a_3 = -\frac{i(\eta - i\mu)}{2\mu}\rho_{11}(\infty), \tag{6.3.24}$$

$$a_4 = \frac{i(\eta + i\mu)}{2\mu}\rho_{11}(\infty). \tag{6.3.25}$$

We note the exponential term with $1/T_2$ of Eq. (6.3.15) does not contribute and we observe the overall factor $\rho_{11}(\infty)$, which turns out to also be expressible as

$$\rho_{11}(\infty) = \frac{\Omega^2}{2(\eta^2 + \mu^2)}. \tag{6.3.26}$$

Thus, we return to Eq. (6.3.1) and find that,

$$g^{(2)}(\tau) = 1 - \frac{i(\eta - i\mu)}{2\mu}e^{-\eta\tau}e^{-i\mu\tau} + \frac{i(\eta + i\mu)}{2\mu}e^{-\eta\tau}e^{+i\mu\tau}, \tag{6.3.27}$$

where we shift to τ for the time variable for compatibility with the literature. We gather similar terms and find

$$g^{(2)}(\tau) = 1 - e^{-\eta\tau}\left(\cos\mu\tau + \frac{\eta}{\mu}\sin\mu\tau\right). \tag{6.3.28}$$

This agrees with Flagg *et al.* (2009, Eq. (2)) when their time variables are all set to τ. In accord with Eq. (6.3.17), we explicitly see

$$g^{(2)}(\tau = 0) = 0. \tag{6.3.29}$$

Cohen-Tannoudji *et al.* (1992, Sec. IV.B.2) show

$$(g^{(2)}(\tau))^* = g^{(2)}(-\tau). \tag{6.3.30}$$

This equation becomes

$$g^{(2)}(\tau) = g^{(2)}(-\tau), \tag{6.3.31}$$

when $g^{(2)}(\tau)$ is real, as it is here. Equation (6.3.31) means calculations must use absolute values of τ for negative τ. Equation (6.3.28) has a dip at $\tau = 0$ due to Eq. (6.3.29). This shows photon anti-bunching, which is explained in Quantum Optics texts such as Fox (2006, Chap. 6) and is illustrated in Figs. 6.5 and 6.6.

The above theoretical developments are now applied to resonance fluorescence experiments that use a quantum dot. As a reminder, resonance fluorescence involves exciting an electron from energy level 0, E_0, to energy level 1, E_1, and detecting the photon emitted when the electron returns to energy level 0. We set

$$\hbar\omega_0 = E_1 - E_0, \qquad (6.3.32)$$

to define ω_0. Generally, a laser is used to excite the electron and the system containing the electron is bathed in an electric field.

Flagg *et al.* (2009) studied quantum dots made of $In_{0.35}Ga_{0.65}As$ within GaAs layers, which are placed inside a planar microcavity constructed by deposited layers. The experiment is done at 10 K to suppress the effects of phonons. The exciting laser is tuned to approximately $\hbar\omega_0 \approx 1.36$ eV or 914 nm and the Rabi energy is of the order of tens of μeV. Thus, the Rabi energy provides a rather fine energy scale, which the experimenters are able to resolve. Further details on the experiment are found in Flagg *et al.* (2009), the Supplementary Information for that paper, and Flagg's Ph.D. thesis (2008). Our focus is on the measurements of $g^{(2)}(\tau)$ by Flagg *et al.* (2009).

We evaluate Eq. (6.3.28) for a range of excitation laser power in order to simulate the trends seen by Flagg *et al.* (2009) for their quantum dot QD1. We reproduce their Fig. 2(d) as the present Fig. 6.7. This gives the relationship between the laser power and the Rabi energy. Since we are at resonance, we directly get the Rabi frequency Ω, which we use in our calculations. T_1 and T_2 are 0.277 and 0.132 ns, respectively, according to Flagg and colleagues. The experimental data for quantum dot QD1 are shown in Fig. 6.8 (Flagg *et al.*, 2009, Fig. 2(b)). We note that $P_0 = 0.05$ mW. The dip centered at $\tau = 0$ is the aforementioned anti-bunching feature. The oscillations in Fig. 6.8 are Rabi oscillations that are distinct from the single peak seen in Fig. 6.6 for a three-level system.

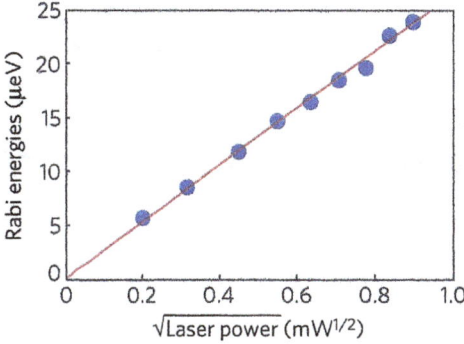

Figure 6.7. The measured relationship between the Rabi energy in μeV and the square root of the laser power in mW, which provides Ω (Flagg *et al.*, 2009, Fig. 2(d)). Adapted by permission from Macmillan Publishers Ltd: Nature Physics, copyright 2009.

We now have the parameters needed for Eqs. (6.3.12) and (6.3.13), and in turn, Eq. (6.3.28). Results are presented for a series of laser powers in Fig. 6.9, which has the results for 0.05, 0.10, 0.20 and 0.40 mW. We find the same trends as the data in Fig. 6.8. The experiments cover both positive and negative τ, but the plots of the calculations only show positive τ. The calculated $g^{(2)}(\tau)$ starts at zero and increases to 1.0 for laser power levels around 0.01 mW, which is not plotted here. Then when the laser power increases, we find $g^{(2)}(\tau)$ starts to exceed and then drops back to 1.0 as the time τ increases. This is the start of a Rabi oscillation. More laser power leads to the clearly discernable Rabi oscillations seen in Figs. 6.8 and 6.9. We remark that reading the periods of the Rabi oscillations from the plots leads to reasonable estimates of the Rabi energies. In addition, our numerical solutions of the Liouville–von Neumann Equations provide $\rho_{11}(\tau)$ that agree with the $\rho_{11}(\tau)$ captured in Eq. (6.3.28). This is a useful consistency check!

We expect $g^{(2)}(\tau)$ to not exceed one for a two-level system. The presence of Rabi oscillations does confuse the situation. Fortunately, here we can relate the period of the oscillations to the Rabi energy. This allows Flagg and his colleagues to postulate that the data come from a two-level system. A glance at Figs. 6.5 and 6.6 shows that they lack evidence of Rabi oscillations. It is likely that the time bin widths

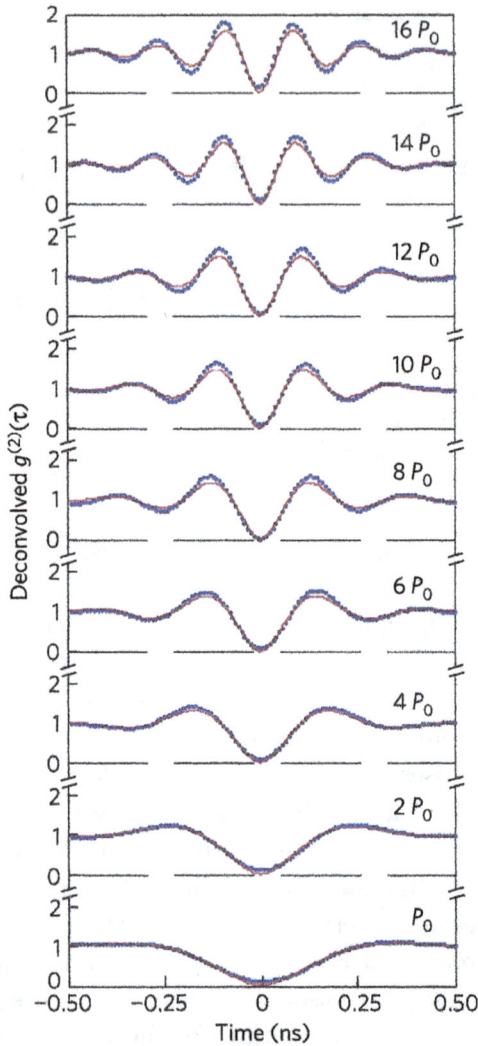

Figure 6.8. The $g^{(2)}(\tau)$ from Flagg *et al.* (2009, Fig. 2(b)). The data are the dots and they have been corrected for the experimental background and the response of their instrument. These data are from their quantum dot QD1. The laser power level $P_0 = 0.05$ mW and the lines are their fits to their equivalent of Eq. (6.3.28). Adapted by permission from Macmillan Publishers Ltd: Nature Physics, copyright 2009.

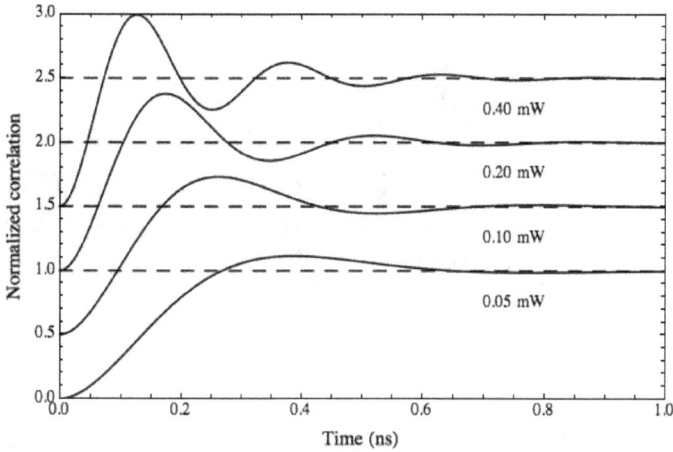

Figure 6.9. Computed $g^{(2)}(\tau)$ from Eq. (6.3.28) for comparison with the data of Flagg *et al.* (2009) shown in Fig. 6.8. The upper 3 plots are offset vertically by 0.5, 1.0 and 1.5, respectively. From bottom to top, the Rabi energies used are 5.5, 8.0, 12.0 and 16.5 μeV, respectively.

are narrow enough to catch Rabi oscillations. Another possibility is the sizes of T_1 and T_2. The values used for Fig. 6.9 are 0.277 and 0.132 ns, respectively. When these are reduced, the Rabi oscillations are reduced and eventually disappear. This behavior is hinted at by Eq. (6.3.28).

The weak-field limit of Eq. (6.3.28) is worked out in Sec. 6.4, which also compares the differential equations of Flagg (2008) with those of Loudon (2000) for the elements of the density matrix.

6.4. The Weak-Field Limit of the Second-Order Correlation Function for a Two-Level System

In Sec. 6.3 we found $g^{(2)}(\tau)$, the second-order correlation function,

$$g^{(2)}(\tau) = 1 - e^{-\eta\tau}\left(\cos\mu\tau + \frac{\eta}{\mu}\sin\mu\tau\right). \qquad (6.4.1)$$

This appears as Eq. (6.3.28) with

$$\eta = \frac{1}{2}\left(\frac{1}{T_1} + \frac{1}{T_2}\right), \qquad (6.4.2)$$

and

$$\mu = \left(\Omega^2 - \frac{1}{4} \left(\frac{1}{T_1} - \frac{1}{T_2} \right)^2 \right)^{1/2}. \qquad (6.4.3)$$

The μ in Eq. (6.4.3) is assumed to be real, so the $g^{(2)}(\tau)$ of Eq. (6.4.1) is appropriate for strong fields. When the field weakens, Ω^2 eventually is less than the other term in the radical for μ. Now μ is replaced by $i\mu$ and we have

$$\cos i\mu\tau = \cosh \mu\tau, \qquad (6.4.4)$$

and

$$\sin i\mu\tau = i \sinh \mu\tau. \qquad (6.4.5)$$

We find

$$g^{(2)}(\tau) = 1 - e^{-\eta\tau} \left(\cosh \mu\tau + \frac{\eta}{\mu} \sinh \mu\tau \right), \qquad (6.4.6)$$

for the weak-field second-order correlation function, which remains real.

This weak-field form differs from a more common expression

$$g^{(2)}(\tau) = (1 - e^{-\tilde{\gamma}_{sp}\tau/2})^2, \qquad (6.4.7)$$

found in Walls (1979) and based on Carmichael and Walls (1976). Here $\tilde{\gamma}_{sp}$ involves spontaneous emission. We derive Eq. (6.4.7) through a comparison of our Liouville–von Neumann Equations for resonance fluorescence in a quantum dot, Eqs. (6.3.2) to (6.3.5), which are based on Flagg (2008), with the equations in Loudon (2000). We rewrite our equations in matrix form

$$\frac{d}{dt} \begin{pmatrix} \rho_{00}(t) \\ \rho_{01}(t) \\ \rho_{10}(t) \\ \rho_{11}(t) \end{pmatrix} = \begin{pmatrix} 0 & \dfrac{i\Omega}{2} & -\dfrac{i\Omega}{2} & \dfrac{1}{T_1} \\ \dfrac{i\Omega}{2} & -\dfrac{1}{T_2} & 0 & -\dfrac{i\Omega}{2} \\ -\dfrac{i\Omega}{2} & 0 & -\dfrac{1}{T_2} & \dfrac{i\Omega}{2} \\ 0 & -\dfrac{i\Omega}{2} & \dfrac{i\Omega}{2} & -\dfrac{1}{T_1} \end{pmatrix} \begin{pmatrix} \rho_{00}(t) \\ \rho_{01}(t) \\ \rho_{10}(t) \\ \rho_{11}(t) \end{pmatrix}, \qquad (6.4.8)$$

and compare them to those in Loudon (2000, Sec. 2.8) with the ρ_{ij} in the order given in Eq. (6.4.8)

$$\frac{d}{dt}\begin{pmatrix} \rho_{00}(t) \\ \rho_{01}(t) \\ \rho_{10}(t) \\ \rho_{11}(t) \end{pmatrix} = \begin{pmatrix} 0 & \dfrac{i\Omega}{2} & -\dfrac{i\Omega}{2} & 2\gamma_{sp} \\ \dfrac{i\Omega}{2} & -\gamma_{sp} & 0 & -\dfrac{i\Omega}{2} \\ -\dfrac{i\Omega}{2} & 0 & -\gamma_{sp} & \dfrac{i\Omega}{2} \\ 0 & -\dfrac{i\Omega}{2} & \dfrac{i\Omega}{2} & -2\gamma_{sp} \end{pmatrix} \begin{pmatrix} \rho_{00}(t) \\ \rho_{01}(t) \\ \rho_{10}(t) \\ \rho_{11}(t) \end{pmatrix}. \quad (6.4.9)$$

In addition, we set per Loudon,

$$\tilde{\gamma}_{sp} = 2\gamma_{sp}. \quad (6.4.10)$$

A comparison of Eqs. (6.4.8) and (6.4.9) leads us to set

$$\frac{1}{T_1} = 2\gamma_{sp}, \quad (6.4.11)$$

and

$$\frac{1}{T_2} = \gamma_{sp}. \quad (6.4.12)$$

These lead to

$$\eta = \frac{3}{2}\gamma_{sp}, \quad (6.4.13)$$

and in the weak-field limit, with the dropping of the Ω^2,

$$\mu = \frac{1}{2}\gamma_{sp}. \quad (6.4.14)$$

Now $g^{(2)}(\tau)$ becomes

$$g^{(2)}(\tau) = 1 - e^{-3\gamma_{sp}\tau/2}\left(\cosh\frac{\gamma_{sp}}{2}\tau + 3\sinh\frac{\gamma_{sp}}{2}\tau\right). \quad (6.4.15)$$

We write the hyperbolic cosh and sinh in terms of exponentials and find

$$g^{(2)}(\tau) = 1 - 2e^{-\gamma_{sp}\tau} + e^{-2\gamma_{sp}\tau} = (1 - e^{-\gamma_{sp}\tau})^2. \quad (6.4.16)$$

When we take Eq. (6.4.10) into account, we see we have found the weak-field limit of Walls (1979).

6.5. What is Next?

We have nibbled at three-level problems in this chapter. Hence, we provide several entry points to such problems after a few closing remarks on two-level systems.

Additional investigations of two-level systems profitably start with Allen and Eberly (1987). Stenholm (2005) covers a variety of solvable two-level systems, as do Bayfield (1999), Tannor (2007) and Berman and Malinovsky (2011). Dubbers and Stöckmann (2013) discuss two-level systems that range from spin precession to neutrino oscillations. Neutrino oscillations are also touched upon by Basdevant and Dalibard (2006) and Pade (2014). Troyan *et al.* (2016) performed a study of an unusual two-level system. They use level splitting in a magnetic field to probe superconductivity in hydrogen sulfide (H_2S). They place a tin (Sn) film enriched with ^{119}Sn in a sample of H_2S. Synchrotron radiation is used to monitor the splitting of the tin energy levels. This level-splitting vanishes when the H_2S sample becomes a superconductor and expels the magnetic field. This is definitely an unexpected use of a two-level system! Magnetic resonance examples may be pursued through Kruk (2016) and Levitt (2008).

A generalization of two-level problems involves replacing the classical electromagnetic field with a quantized field. The Jaynes–Cummings Model (1963) is a useful first step and is discussed by Gerry and Knight (2005). Braak (2011) leaped beyond this and found an analytical solution to the Quantum Rabi Model. Braak (2016) extends his methods to further models in Quantum Optics. Quantized fields are also used in Carmichael (1999, Chap. 2) to derive the spectrum of the Mollow (1969) triplets. These occur when a two-level system is subjected to a strong electric field.

I treat monochromatic fields in this book, but many experiments use pulses. Grossmann (2008) and Shore (2011) are helpful in understanding these added challenges. Pulse shapes and sequences are under investigation for controlling the energy of emission and absorption of a two-level system. Fotso and Dobrovitski (2017) is an example of such studies, as well as a source for earlier work.

The addition of a third level takes us to Scully and Zubairy (1997), Bayfield (1999), Loudon (2000), Stenholm (2005), Lambropoulos and

Petrosyan (2007), Shore (2011), Berman and Malinovsky (2011), and van der Straten and Metcalf (2016). All of these references develop the formalism and treat situations such as dark states and Raman scattering. Scully and Zubairy, Shore, and van der Straten and Metcalf, in particular, explain these two topics. Finally, the two volumes by Shore (1990) should be consulted for illuminations of few-level systems.

With these pointers, I rest my pen.

References

G. S. Agarwal, *Quantum Optics* (Cambridge University Press, Cambridge, 2013).

I. Aharonovich, S. Castelletto, D. A. Simpson, C.-H. Su, A. D. Greentree, and S. Prawer, "Diamond-based single-photon emitters", *Reports on Progress in Physics* **74**, 076501 (2011).

L. Allen and J. H. Eberly, *Optical Resonance and Two-Level Atoms* (Dover Publications, New York, 1987).

T. Basché, S. Kummer, and C. Bräuchle, "Excitation and Emission Spectroscopy and Quantum Optical Measurements", in *Single-Molecule Optical Detection, Imaging and Spectroscopy*, T. Basché, W. E. Moerner, M. Orrit, and U. P. Wild, eds. (VCH Verlags-gesellschaft, Weinheim, 1997), pp. 31–67.

J.-L. Basdevant and J. Dalibard, *The Quantum Mechanics Solver: How to Apply Quantum Theory to Modern Physics*, 2$^\text{nd}$ ed. (Springer-Verlag, Berlin, 2006).

J. E. Bayfield, *Quantum Evolution: An Introduction to Time-Dependent Quantum Mechanics* (John Wiley and Sons, New York, 1999).

J. C. Bergquist, R. G. Hulet, W. M. Itano, and D. J. Wineland, "Observation of quantum jumps in a single atom", *Phys. Rev. Lett.* **57**, 1699–1702 (1986).

P. R. Berman and V. S. Malinovsky, *Principles of Laser Spectroscopy and Quantum Optics* (Princeton University Press, Princeton, 2011).

D. Braak, "Integrability of the Rabi Model", *Phys. Rev. Lett.* **107**, 100401 (2011).

D. Braak, "Analytical Solutions of Basic Models in Quantum Optics", in *Applications + Practical Conceptualization + Mathematics = Fruitful Innovation*, Mathematics for Industry Vol. 11, R. S. Anderssen, P. Broadbridge, Y. Fukumoto, K. Kajiwara, T. Tagaki, E. Verbitskiy, and M. Wakayama, eds. (Springer Japan, 2016), pp. 75–92.

H. J. Carmichael, *Statistical Methods in Quantum Optics 1* (Springer-Verlag, Berlin, 1999).

H. J. Carmichael and D. F. Walls, "A quantum-mechanical master equation treatment of the dynamical Stark effect", *J. Phys. B: Atom. Molec. Phys.* **9**, 1199–1219 (1976).

C. Cohen-Tannoudji, J. Dupont-Roc, and G. Grynberg, *Atom-Photon Interactions: Basic Processes and Applications* (John Wiley and Sons, New York, 1992).

X. Ding, Y. He, Z.-C. Duan, N. Gregersen, M.-C. Chen, S. Unsleber, S. Maier, C. Schneider, M. Kemp, S. Höfling, C.-Y. Lu, and J.-W. Pan, "On-demand single photons with high extraction efficiency and near-unity indistinguishability from a resonantly driven quantum dot in a micropillar", *Phys. Rev. Lett.* **116**, 020401 (2016).

D. Dubbers and H.-J. Stöckmann, *Quantum Physics: The Bottom-Up Approach* (Springer-Verlag, Berlin, 2013).

E. B. Flagg, II, "Coherent control and decoherence of single semiconductor quantum dots in a microcavity", PhD thesis, The University of Texas at Austin (2008).

E. B. Flagg, A. Muller, J. W. Robertson, S. Founta, D. G. Deppe, M. Xiao, W. Ma, G. J. Salamo, and C. K. Shih, "Resonantly driven coherent oscillations in a solid-state quantum emitter", *Nature Physics* **5**, 203–207, (2009).

H. F. Fotso and V. V. Dobrovitski, "Absorption spectrum of a two-level system subjected to a periodic pulse sequence", *Phys. Rev. B* **95**, 214301 (2017).

M. Fox, *Quantum Optics: An Introduction* (Oxford University Press, Oxford, 2006), reprinted 2007.

C. Galland, Y. Ghosh, A. Steinbrück, M. Sykora, J. A. Hollingsworth, V. I. Klimov, and H. Htoon, "Two types of luminescence blinking revealed by spectroelectrochemistry of single quantum dots", *Nature* **479**, 203–208 (2011).

C. C. Gerry and P. L. Knight, *Introductory Quantum Optics* (Cambridge University Press, Cambridge, 2004).

F. Grossmann, *Theoretical Femtosecond Physics: Atoms and Molecules in Strong Laser Fields* (Springer, Berlin, 2008).

E. T. Jaynes and F. W. Cummings, "Comparison of quantum and semiclassical radiation theories with application to the beam maser", *Proceedings of the IEEE* **51**, 89–109 (1963).

D. Kruk, *Understanding Spin Dynamics* (Pan Stanford Publishing, Singapore, 2016).

A. V. Kuhlmann, J. H. Prechtel, J. Houel, A. Ludwig, D. Reuter, A. D. Wieck, and R. J. Warburton, "Transform-limited single photons from a single quantum dot", *Nature Communications* **6**, 9204 (2015).

M. Kuno, D. P. Fromm, H. F. Hamann, A. Gallagher, and D. J. Nesbitt, "'On'/'Off' fluorescence intermittency of single semiconductor quantum dots", *J. Chem. Phys.* **115**, 1028–1040 (2001).

P. Lambropoulos and D. Petrosyan, *Fundamentals of Quantum Optics and Quantum Information* (Springer-Verlag, Berlin, 2007).

M. H. Levitt, *Spin Dynamics: Basics of Nuclear Magnetic Resonance*, 2$^\text{nd}$ ed. (John Wiley and Sons, Chichester, 2008).

R. Loudon, *The Quantum Theory of Light*, 3$^\text{rd}$ ed. (Oxford University Press, Oxford, 2000).

Mathematica = Wolfram Research, Inc., Mathematica, Version 9.0.0.0, Champaign, IL (2012).

A. Migdal, S. V. Polyakov, J. Fan, and J. C. Bienfang, eds., *Single-Photon Generation and Detection: Physics and Applications*, Experimental Methods in the Physical Sciences, Vol. 45 (Academic Press, Amsterdam, 2013).

B. R. Mollow, "Power spectrum of light scattered by two-level systems", *Phys. Rev.* **188**, 1969–1975 (1969).

W. Nagourney, J. Sandberg, and H. Dehmelt, "Shelved optical electron amplifier: observation of quantum jumps", *Phys. Rev. Lett.* **56**, 2797–2799 (1986).

I. S. Osad'ko, *Selective Spectroscopy of Single Molecules* (Springer, Berlin, 2003).

J. Pade, *Quantum Mechanics for Pedestrians 1: Fundamentals* (Springer International Publishing, Cham, 2014).

M.-E. Pistol, P. Castrillo, D. Hessman, J. A. Prieto, and L. Samuelson, "Random telegraph noise in photoluminescence from individual self-assembled quantum dots", *Phys. Rev B* **59**, 10725–10729 (1999).

A. Predojević and M. W. Mitchell, eds., *Engineering the Atom-Photon Interaction: Controlling Fundamental Processes with Photons, Atoms and Solids* (Springer International Publishing, Cham, Switzerland, 2015).

T. Sauter, W. Neuhauser, R. Blatt, and P. E. Toschek, "Observation of Quantum Jumps", *Phys. Rev. Lett.* **57**, 1696–1698 (1986).

M. O. Scully and M. Suhail Zubairy, *Quantum Optics* (Cambridge University Press, Cambridge, 1997).

B. W. Shore, *Manipulating Quantum Structures Using Laser Pulses* (Cambridge University Press, Cambridge, 2011).

B. W. Shore, *The Theory of Coherent Atomic Excitation*, Vols. 1 and 2 (John Wiley and Sons, New York, 1990).

D. A. Simpson, E. Ampem-Lassen, B. C. Gibson, S. Trpkovski, F. M. Hossain, S. T. Huntington, A. D. Greentree, L. C. L. Hollenberg, and S. Prawer, "A highly efficient two level diamond based single photon source", *Appl. Phys. Lett.* **94**, 203107 (2009).

F. D. Stefani, J. P. Hoogenboom, and E. Barkai, "Beyond quantum jumps: Blinking nanoscale light emitters", *Physics Today* **62** (February), 34–39 (2009).

S. Stenholm, *Foundations of Laser Spectroscopy* (Dover Publications, Mineola, New York, 2005).

D. J. Tannor, *Introduction to Quantum Mechanics: A Time-Dependent Perspective* (University Science Books, Sausalito, California, 2007).

I. Troyan, A. Gavriliuk, R. Rüffer, A. Chumakov, A. Mironovich, I. Lyubutin, D. Perekalin, A. P. Drozdov, and M. I. Eremets, "Observation of superconductivity in hydrogen sulfide from nuclear resonant scattering", *Science* **351**, 1303–1306 (2016).

P. van der Straten and H. Metcalf, *Atoms and Molecules Interacting with Light: Atomic Physics for the Laser Era* (Cambridge University Press, Cambridge, 2016).

D. F. Walls, "Evidence for the quantum nature of light", *Nature* **280**, 451–454 (1979).

Appendices

Appendix 1. Physical Constants

The following values for various physical constants and particle properties come from http://physics.nist.gov/cuu/Constants/ and are the 2014 values. The site was accessed on 11 and 22 May 2016, and 10 March 2017.

Planck's constant	6.626070×10^{-34} joule second
Planck's constant / 2π	$1.0545718 \times 10^{-34}$ joule second
Planck's constant / 2π	$6.58211951 \times 10^{-16}$ eV second
Speed of light in vacuum	299792458 m/s
Elementary charge	$1.6021766 \times 10^{-19}$ coulombs
Boltzmann's constant	$1.3806485 \times 10^{-23}$ joule/kelvin
Electron mass	9.109383×10^{-31} kg
Electron g factor	-2.002319304361
Bohr magneton	$9.27400999 \times 10^{-24}$ J/T
Muon mass	$1.883531594 \times 10^{-28}$ kg
Muon g factor	-2.0023318418
Neutron mass	$1.674927471 \times 10^{-27}$ kg
Neutron g factor	-3.82608545
Proton mass	$1.672621898 \times 10^{-27}$ kg
Proton g factor	5.585694702
Nuclear magneton	$5.050783699 \times 10^{-27}$ J/T
Vacuum permittivity ε_0	$8.854187817 \times 10^{-12}$ farad/m
Vacuum permeability μ_0	$4\pi \times 10^{-7}$ N/A^2 (1 N/A^2 = 1 henry/m)

Appendix 2. The Electromagnetic Field

It is useful to define several quantities associated with electromagnetic fields, since we treat many time-dependent problems that involve such fields. Quantities such as the energy density also frequently appear in descriptions of experiments. We use MKS units and, for clarity, plane waves are used here.

Plane waves have spatial and temporal dependences and are specified by the vectors

$$\bar{E}(\bar{r}, t) = E_0(\omega)\bar{\varepsilon}\sin(\bar{k} \cdot \bar{r} - \omega t + \alpha), \tag{A.2.1}$$

$$\bar{B}(\bar{r}, t) = E_0(\omega)\frac{1}{\omega}(\bar{k} \times \bar{\varepsilon})\sin(\bar{k} \cdot \bar{r} - \omega t + \alpha). \tag{A.2.2}$$

Here $\bar{E}$ is the electric field and $\bar{B}$ is the magnetic field. The vector $\bar{r}$ has the spatial coordinates and t is the time. The angular frequency of this monochromatic plane wave is $\omega = 2\pi\nu$ with ν the frequency. The electric field is in the direction of the unit polarization vector $\bar{\varepsilon}$, which is perpendicular to the direction $\bar{k}$ of the wave's propagation. The magnitude of $\bar{k}$ is $2\pi/\lambda$ where the wavelength λ is defined by

$$\lambda\nu = c, \tag{A.2.3}$$

with c the speed of light in vacuum. Finally, α is a phase shift. One can show that the above $\bar{E}$ and $\bar{B}$ satisfy Maxwell's Equations.

Equations (A.2.1) and (A.2.2) specify a linearly-polarized electromagnetic field. There are two independent polarizations that are perpendicular to $\bar{k}$. Other polarizations may be represented by a linear superposition of these two polarizations. An additional superposition over frequency allows us to represent still more general electromagnetic fields.

We next relate the energy density of the electromagnetic field to the photon density. Each photon has energy $h\nu$ with h equal to Planck's constant. We note

$$h\nu = \hbar\omega, \tag{A.2.4}$$

with $\hbar = h/2\pi$. We use both expressions for the photon's energy. In addition, we introduce the dielectric permittivity ε_0 and the magnetic permeability μ_0 of the vacuum. These are related by

$$\varepsilon_0\mu_0 = 1/c^2. \tag{A.2.5}$$

The energy density $U(\omega)$ of the electromagnetic field is

$$U(\omega) = \frac{1}{2}\left(\varepsilon_0|\bar{E}|^2 + \frac{1}{\mu_0}|\bar{B}|^2\right), \tag{A.2.6}$$

so we need $|\bar{B}|^2$ and this involves

$$|\bar{k} \times \bar{\varepsilon}| = |\bar{k}||\bar{\varepsilon}|\sin\theta. \tag{A.2.7}$$

Here θ is the angle between $\bar{k}$ and $\bar{\varepsilon}$ and we have $\theta = \pi/2$ because the two vectors are perpendicular. Thus, with $\bar{\varepsilon}$ a unit vector,

$$|\bar{k} \times \bar{\varepsilon}| = |\bar{k}| = 2\pi/\lambda. \tag{A.2.8}$$

We now evaluate the energy density, Eq. (A.2.6). With the help of Eqs. (A.2.2) and (A.2.8), we see the magnetic field term of Eq. (A.2.6) involves

$$\frac{1}{\mu_0}|E_0|^2\frac{1}{\omega^2}|\bar{k}|^2 = \frac{1}{\mu_0}|E_0|^2\frac{1}{(2\pi\nu)^2}\left(\frac{2\pi}{\lambda}\right)^2$$

$$= \frac{1}{\mu_0}|E_0|^2\left(\frac{1}{c^2}\right) = \varepsilon_0|E_0|^2, \tag{A.2.9}$$

and we have used Eq. (A.2.5) also. This leads to

$$U(\omega) = \frac{1}{2}(\varepsilon_0|E_0|^2 + \varepsilon_0|E_0|^2)\sin^2(\bar{k}\cdot\bar{r} - \omega t + \alpha)$$

$$= \varepsilon_0|E_0|^2\sin^2(\bar{k}\cdot\bar{r} - \omega t + \alpha). \tag{A.2.10}$$

We note E_0 may depend on ω and we average the energy density over a period

$$T = 2\pi/\omega = 2\pi/(2\pi\nu) = 1/\nu. \tag{A.2.11}$$

This average of the energy density over a period requires

$$\frac{1}{T}\int_0^T dt\, \sin^2(\bar{k}\cdot\bar{r} - \omega t + \alpha) = \frac{1}{2T}\int_0^T dt\{1 - \cos 2(\bar{k}\cdot\bar{r} - \omega t + \alpha)\}$$

$$= \frac{1}{2} + \frac{1}{4\omega T}[\sin 2(\bar{k}\cdot\bar{r} + \alpha)\cos 2\omega t - \cos 2(\bar{k}\cdot\bar{r} + \alpha)\sin 2\omega t]_0^T$$

$$= \frac{1}{2} + \frac{1}{4\omega T}[\sin 2(\bar{k}\cdot\bar{r} + \alpha) - \sin 2(\bar{k}\cdot\bar{r} + \alpha)] = \frac{1}{2}. \tag{A.2.12}$$

We go from the third expression to the fourth since

$$2\omega T = 2(2\pi\nu)/\nu = 4\pi. \tag{A.2.13}$$

Hence, we have $\cos 4\pi = 1$, $\cos 0 = 1$, $\sin 4\pi = 0$ and $\sin 0 = 0$. Thus, we come to

$$\frac{1}{T}\int_0^T dt\, U(\omega) = \frac{1}{2}\varepsilon_0 |E_0(\omega)|^2 = \rho(\omega), \tag{A.2.14}$$

and we have introduced the time-averaged energy density $\rho(\omega)$.

Photons lead us to a second expression for the time-averaged energy density. Let $N(\omega)$ be the number of photons of angular frequency ω within a volume V. Then we express

$$\rho(\omega) = \hbar\omega N(\omega)/V. \tag{A.2.15}$$

When we set these two expressions for the time-averaged energy density equal, we find

$$|E_0(\omega)| = [2\hbar\omega N(\omega)/(\varepsilon_0 V)]^{1/2}. \tag{A.2.16}$$

This gives us a relationship between the magnitude of the electric field and the photon expressions. The intensity $I(\omega)$ is the average rate of energy flow through a unit area perpendicular to the direction of propagation $\bar{k}$. Then with Eq. (A.2.15)

$$I(\omega) = \rho(\omega)c = \hbar\omega c N(\omega)/V. \tag{A.2.17}$$

In matter, c is replaced by c/n where n is the index of refraction for the material, and n may be dependent on the angular frequency ω.

We close this appendix with the Poynting vector in vacuum,

$$\bar{S} = \frac{1}{\mu_0}\bar{E} \times \bar{B}. \tag{A.2.18}$$

Here $\bar{S}$ is the rate of energy transport per unit area, with units of joules/(s-m^2). We let E_0 and B_0 be the magnitudes of the electric and magnetic fields, respectively. For plane waves, Maxwell's Equations lead to

$$B_0 = E_0/c, \tag{A.2.19}$$

and the Poynting vector time-averaged over a period is

$$S = \frac{1}{2}E_0^2/(c\mu_0). \tag{A.2.20}$$

Further developments are found in Bransden and Joachain (2000) and in most electricity and magnetism texts such as Jackson (1999) and Band (2006).

References

Y. B. Band, *Light and Matter: Electromagnetism, Optics, Spectroscopy and Lasers* (John Wiley and Sons, Ltd, Chicester, 2006).

B. H. Bransden and C. J. Joachain, *Quantum Mechanics*, 2nd ed. (Prentice Hall, Harlow, England, 2000).

J. D. Jackson, *Classical Electrodynamics*, 3rd ed. (John Wiley and Sons, New York, 1999).

Appendix 3. Diagonalization of a Matrix in Chapter 2

We need to show that the matrix M of Eq. (2.2.27) diagonalizes the matrix

$$A = \begin{pmatrix} -\Delta & V \\ V & +\Delta \end{pmatrix}, \tag{A.3.1}$$

from Eq. (2.2.9). We factor out Δ, which will be returned later, and use Eq. (2.2.20)

$$\tan 2\theta = V/\Delta. \tag{A.3.2}$$

This allows us to write

$$\sin 2\theta = V/(\Delta^2 + V^2)^{1/2}$$
$$\cos 2\theta = \Delta/(\Delta^2 + V^2)^{1/2}. \tag{A.3.3}$$

We next check that Eq. (2.2.28) provides the inverse to M,

$$
\begin{aligned}
MM^{-1} &= \begin{pmatrix} \cos\theta & -\sin\theta \\ \sin\theta & \cos\theta \end{pmatrix} \begin{pmatrix} \cos\theta & \sin\theta \\ -\sin\theta & \cos\theta \end{pmatrix} \\
&= \begin{pmatrix} \cos^2\theta + \sin^2\theta & \cos\theta\sin\theta - \sin\theta\cos\theta \\ \sin\theta\cos\theta - \cos\theta\sin\theta & \sin^2\theta + \cos^2\theta \end{pmatrix} \\
&= \begin{pmatrix} 1 & 0 \\ 0 & 1 \end{pmatrix}.
\end{aligned} \tag{A.3.4}
$$

Hence, M^{-1} is the inverse of M.

We now diagonalize A by evaluating MAM^{-1}

$$\begin{pmatrix} \cos\theta - \sin\theta \\ \sin\theta \ \ \cos\theta \end{pmatrix} \begin{pmatrix} -1 \ \ \tan 2\theta \\ \tan 2\theta \ \ 1 \end{pmatrix} \begin{pmatrix} \cos\theta \ \ \sin\theta \\ -\sin\theta \ \cos\theta \end{pmatrix} = \begin{pmatrix} \cos\theta - \sin\theta \\ \sin\theta \ \ \cos\theta \end{pmatrix}$$

$$\times \begin{pmatrix} -\cos\theta - \tan 2\theta \sin\theta \ \ -\sin\theta + \tan 2\theta \cos\theta \\ \tan 2\theta \cos\theta - \sin\theta \ \ \ \ \tan 2\theta \sin\theta + \cos\theta \end{pmatrix}. \tag{A.3.5}$$

We let D be the matrix product in Eq. (A.3.5) and go term by term.
First,

$$D_{11} = -\cos^2\theta - \cos\theta \tan 2\theta \sin\theta - \sin\theta \tan 2\theta \cos\theta + \sin^2\theta$$

$$= -(\cos^2\theta - \sin^2\theta) - 2\sin\theta\cos\theta\tan 2\theta$$

$$= -\cos 2\theta - \sin 2\theta \tan 2\theta. \tag{A.3.6}$$

We now bring back Δ and use Eqs. (A.3.2) and (A.3.3) to find

$$\Delta D_{11} = -(\Delta^2 + V^2)/(\Delta^2 + V^2)^{1/2} = -(\Delta^2 + V^2)^{1/2}, \tag{A.3.7}$$

and this agrees with Eq. (2.2.4).
We next discover

$$D_{12} = -2\sin\theta\cos\theta + \tan 2\theta(\cos^2\theta - \sin^2\theta)$$

$$= -\sin 2\theta + \tan 2\theta \cos 2\theta = 0, \tag{A.3.8}$$

and

$$D_{21} = D_{12} = 0. \tag{A.3.9}$$

Finally,

$$D_{22} = \cos^2\theta - \sin^2\theta + 2\sin\theta\cos\theta\tan 2\theta = \cos 2\theta + \sin 2\theta \tan 2\theta. \tag{A.3.10}$$

This leads to

$$\Delta D_{22} = (\Delta^2 + V^2)/(\Delta^2 + V^2)^{1/2} = (\Delta^2 + V^2)^{1/2}, \tag{A.3.11}$$

and this also agrees with Eq. (2.2.4).

Thus, we have shown that the matrix M does, indeed, diagonalize the matrix A of Eq. (A.3.1).

Appendix 4. The Time Evolution Operator and the Sudden Approximation

We wish to use the time evolution operator to find a condition for the validity of the Sudden Approximation. We build on the discussion of Duffey (1992, pp. 254–256). Our Hamiltonian $\hat{H}_1$ is independent of the time t. We define

$$\hat{H} = \hat{H}_1 \quad \text{for} \quad t < 0, \tag{A.4.1}$$

$$\hat{H}(t) = \hat{H}_1 + \hat{U}(t) \quad \text{for} \quad 0 < t < \Delta t, \tag{A.4.2}$$

and

$$\hat{H} = \hat{H}_1 + \hat{U}(\Delta t) = \hat{H}_2 \quad \text{for} \quad t > \Delta t. \tag{A.4.3}$$

We note $\hat{H}_2$ is independent of time. The Hamiltonian starts as $\hat{H}_1$ and then transitions to $\hat{H}_2$ over a time interval of Δt.

The wave function for $t > \Delta t$ is written as

$$\psi(\bar{r}, t) = e^{-i\hat{H}_2(t-\Delta t)/\hbar} \psi(\bar{r}, \Delta t), \tag{A.4.4}$$

in agreement with Sec. 1.2. We now express the wave function at Δt in terms of that at $t = 0$

$$\psi(\bar{r}, \Delta t) = \exp\left\{-i\left[\hat{H}_1 \Delta t + \int_0^{\Delta t} \hat{U}(t')dt'\right]/\hbar\right\} \psi(\bar{r}, 0). \tag{A.4.5}$$

And in accord with Sakurai and Napolitano (2011, Sec. 2.1), we assume the Hamiltonians at different times commute. We further assume that the components of $\hat{H}(t)$ in Eq. (A.4.2) commute. Then we put Eq. (A.4.5) in Eq. (A.4.4) with Eq. (A.4.3) to find that the argument of the exponential becomes

$$-\frac{i}{\hbar}\left[\hat{H}_1(t-\Delta t)+\hat{U}(\Delta t)(t-\Delta t)+\hat{H}_1\Delta t+\int_0^{\Delta t}\hat{U}(t')dt'\right].$$

(A.4.6)

This simplifies, slightly, to

$$-\frac{i}{\hbar}\left[\hat{H}_1 t+\hat{U}(\Delta t)t-\hat{U}(\Delta t)[\Delta t]+\int_0^{\Delta t}\hat{U}(t')dt'\right].$$
(A.4.7)

The first two terms help the wave function evolve from $t=\Delta t$. It is useful to assume a form for $\hat{U}(t)$ and we take

$$\hat{U}(t)=\alpha t,$$
(A.4.8)

which allows us to do the integral in Eq. (A.4.7). The result for terms 3 and 4 of Eq. (A.4.7) is

$$-\alpha(\Delta t)^2+\frac{\alpha}{2}(\Delta t)^2=-\frac{\alpha}{2}(\Delta t)^2=-\frac{1}{2}\hat{U}(\Delta t)\Delta t.$$
(A.4.9)

We look back and see if

$$\frac{1}{2}|\hat{U}(\Delta t)|\Delta t\ll\hbar,$$
(A.4.10)

then the Sudden Approximation is valid and we have

$$\psi(\bar{r},t)=\exp\left\{-\frac{i}{\hbar}[(\hat{H}_1+\hat{U}(\Delta t))t]\right\}\psi(\bar{r},0).$$
(A.4.11)

We find the wave function evolves according to $\hat{H}_2$ of Eq. (A.4.3).

References

G. H. Duffey, *Quantum States and Processes* (Prentice Hall, Englewood Cliffs, N. J., 1992).

J. J. Sakurai and J. Napolitano, *Modern Quantum Mechanics*, 2nd ed. (Addison-Wesley, Boston, 2011).

Appendix 5. Beyond the Rotating Wave Approximation

We explore the differences between the Rotating Wave Approximation (RWA) and the full solution to the coupled, first-order differential equations for the two-level problem. Section 3.3.1 shows that the RWA solution is sometimes quite close to the full solution. Here we delve into how the high-frequency components of the full solution behave, and we draw upon Secs. 3.2 and 3.3.1. We reveal how the high-frequency oscillations decrease when $\omega + \omega_0$ increases.

We start with the equations for the full solution,

$$\frac{dc_1(t)}{dt} = (i\mu E/2\hbar)(e^{i(\omega-\omega_0)t} + e^{-i(\omega+\omega_0)t})c_2(t), \qquad \text{(A.5.1)}$$

$$\frac{dc_2(t)}{dt} = (i\mu E/2\hbar)(e^{-i(\omega-\omega_0)t} + e^{+i(\omega+\omega_0)t})c_1(t). \qquad \text{(A.5.2)}$$

Some notation is helpful. Let

$$\tilde{\Omega} = \mu E/\hbar, \qquad \text{(A.5.3)}$$

$$\delta\omega = \omega - \omega_0 = \omega - (\omega_2 - \omega_1), \qquad \text{(A.5.4)}$$

with the ground state energy level 1 and the excited state energy level 2. Next, we recast Eqs. (A.5.1) and (A.5.2) in terms of dimensionless variables. We start with the time

$$\tau = \tilde{\Omega}t, \qquad \text{(A.5.5)}$$

and go on to angular frequencies

$$\delta = (\omega - \omega_0)/\tilde{\Omega}, \qquad \text{(A.5.6)}$$

$$\gamma = (\omega + \omega_0)/\tilde{\Omega}. \qquad \text{(A.5.7)}$$

Equations (A.5.1) and (A.5.2) become

$$\frac{dc_1(\tau)}{d\tau} = (i/2)(e^{i\delta\tau} + e^{-i\gamma\tau})c_2(\tau), \qquad (A.5.8)$$

and

$$\frac{dc_2(\tau)}{d\tau} = (i/2)(e^{-i\delta\tau} + e^{i\gamma\tau})c_1(\tau). \qquad (A.5.9)$$

Figures 3.6 to 3.8 show how the high-frequency components of the full solution for resonance decrease in amplitude when γ is greater than 1 and is increasing. We demonstrate here how the high-frequency components go as $1/\gamma$. We follow the ideas of Bonacci (2004) although our notation is slightly different, as is our argument. Bonacci points out if the equations for c_1 and c_2 are recast as second-order ordinary differential equations, then we discover terms with coefficients of γ and δ. We now derive these equations.

We differentiate Eq. (A.5.8) with respect to τ and find

$$\frac{d^2c_1}{d\tau^2} = \frac{i}{2}(i\delta e^{i\delta\tau} - i\gamma e^{-i\gamma\tau})c_2(\tau) + \frac{i}{2}(e^{i\delta\tau} + e^{-i\gamma\tau})\frac{d}{d\tau}c_2(\tau).$$
$$(A.5.10)$$

When Eq. (A.5.9) is substituted into Eq. (A.5.10), the result is

$$\frac{d^2c_1}{d\tau^2} = \frac{1}{2}(-\delta e^{i\delta\tau} + \gamma e^{-i\gamma\tau})c_2(\tau)$$
$$- \frac{1}{4}(2 + e^{-i(\delta+\gamma)\tau} + e^{i(\delta+\gamma)\tau})c_1(\tau), \qquad (A.5.11)$$

which simplifies to

$$\frac{d^2c_1}{d\tau^2} = \frac{1}{2}(-\delta e^{i\delta\tau} + \gamma e^{-i\gamma\tau})c_2(\tau) - \frac{1}{2}(1 + \cos((\delta+\gamma)\tau))c_1(\tau).$$
$$(A.5.12)$$

This shows the terms with coefficients of γ and δ as well as an exponential and a cosine with high-frequency arguments. A parallel treatment of Eq. (A.5.9) yields

$$\frac{d^2c_2}{d\tau^2} = \frac{1}{2}(\delta e^{-i\delta\tau} - \gamma e^{i\gamma\tau})c_1(\tau) - \frac{1}{2}(1 + \cos((\delta+\gamma)\tau))c_2(\tau).$$
$$(A.5.13)$$

We assume $\delta = 0$, so our argument applies to resonance. We work with c_1 and we note that c_2 leads to the same conclusions. Our starting point is Eq. (A.5.12) with $\delta = 0$

$$\frac{d^2 c_1}{d\tau^2} = \frac{1}{2}(\gamma e^{-i\gamma\tau})c_2(\tau) - \frac{1}{2}(1 + \cos(\gamma\tau))c_1(\tau). \qquad (A.5.14)$$

Both the exponential term and the cosine term show high-frequency oscillations, but we keep only the former, because its coefficient is γ and we assume $\gamma \gg 1$. Thus, we seek a solution of

$$\frac{d^2 c_1}{d\tau^2} = \frac{1}{2}(\gamma e^{-i\gamma\tau})c_2(\tau) - \frac{1}{2}c_1(\tau). \qquad (A.5.15)$$

We break the solution into two pieces with

$$c_1(\tau) = \check{c}_1(\tau) + \tilde{c}_1(\tau), \qquad (A.5.16)$$

and assume a similar decomposition for $c_2(\tau)$.

In analogy with the RWA, we drop the high-frequency term and first solve

$$\frac{d^2 \check{c}_1}{d\tau^2} = -\frac{1}{2}\check{c}_1(\tau), \qquad (A.5.17)$$

and the analogous equation for $\check{c}_2(\tau)$ with

$$\check{c}_1(\tau) = \cos(\tau/\sqrt{2}), \qquad (A.5.18)$$

and

$$\check{c}_2(\tau) = \sin(\tau/\sqrt{2}). \qquad (A.5.19)$$

We note we are applying the initial conditions

$$\check{c}_1(0) = 1 \quad \text{and} \quad \check{c}_2(0) = 0. \qquad (A.5.20)$$

Our expectation is that the $\tilde{c}_i(\tau)$ may be treated as a perturbation upon the $\check{c}_i(\tau)$. We substitute Eq. (A.5.16) into Eq. (A.5.15), keep the leading terms and use Eq. (A.5.19) to find

$$\frac{d^2\tilde{c}_1}{d\tau^2} = \frac{1}{2}(\gamma e^{-i\gamma\tau})\sin(\tau/\sqrt{2}). \tag{A.5.21}$$

We use the Integrate routine of Mathematica and learn that the high-frequency term is

$$\tilde{c}_1(\tau) = \frac{\gamma}{2(2\gamma^2 - 1)}\left\{\frac{e^{-i\gamma\tau}}{2\gamma^2 - 1}Q(\tau)\right\} - \frac{4\sqrt{2}i\gamma^2}{2[2\gamma^2 - 1]^2} - \frac{\sqrt{2}\gamma\tau}{2(2\gamma^2 - 1)}, \tag{A.5.22}$$

with

$$Q(\tau) = 4\sqrt{2}i\gamma\cos(\tau/\sqrt{2}) - 2(2\gamma^2 + 1)\sin(\tau/\sqrt{2}). \tag{A.5.23}$$

We note that $\tilde{c}_1(0) = 0$. An examination of $\tilde{c}_i(\tau)$ shows that all the terms decrease at least as $1/\gamma$, when γ increases. Thus, we see in this limit

$$\tilde{c}_1(\tau) \propto \frac{1}{\gamma}, \tag{A.5.24}$$

and the high-frequency oscillations in the full solution for $c_1(\tau)$ decrease as $1/\gamma$. A similar behavior is found for $\tilde{c}_2(\tau)$. (The last term on the right-hand side of Eq. (A.5.22) is linear in τ, which is disconcerting, and serves to remind us we are using a perturbation approach.)

A more direct approach is possible. We assume resonance and approximate the first rise and fall of the full solution for the occupation probability $p_2(t)$ with

$$(1 - q)(\sin\tilde{\Omega}t/2)^2 + q(\sin\omega t)^2. \tag{A.5.25}$$

We find that as ω increases the factor q decreases and this reduces the amplitude of the high-frequency term, which is the second term in Eq. (A.5.25).

Figures 3.6 to 3.8 show that the amplitude of the high-frequency oscillations do decrease with the increase of ω. We close this Appendix with three more examples of the full solutions versus the RWA solutions. We set $\tilde{\Omega} = 0.5$ and have $\omega = \omega_0 = 4$ in Fig. A5.1, $\omega = \omega_0 = 6$ in Fig. A5.2, and $\omega = \omega_0 = 8$ in Fig. A5.3. The γ are

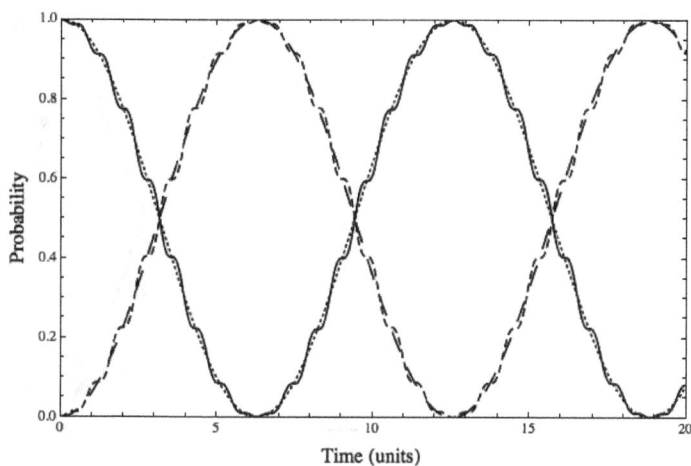

Figure A5.1. Occupation probabilities for a two-level system with $\omega = \omega_0 = 4$ and $\tilde{\Omega} = 0.5$. The system starts in level 1, the lower energy level. $p_1(t)$ is solid and $p_2(t)$ is short-dashed for the full solution. $p_1(t)$ is dotted and $p_2(t)$ is long-dashed for the RWA solution.

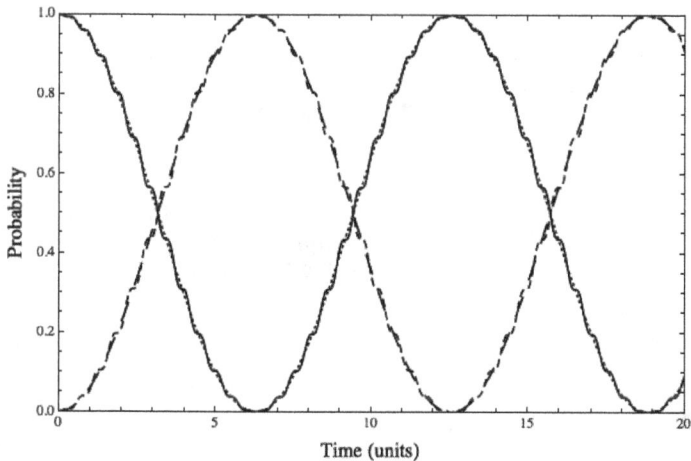

Figure A5.2. Occupation probabilities for a two-level system with $\omega = \omega_0 = 6$ and $\tilde{\Omega} = 0.5$. The system starts in level 1, the lower energy level. $p_1(t)$ is solid and $p_2(t)$ is short-dashed for the full solution. $p_1(t)$ is dotted and $p_2(t)$ is long-dashed for the RWA solution.

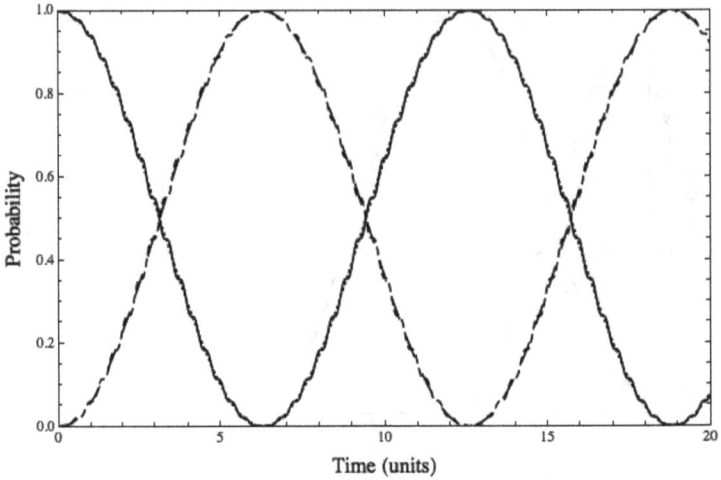

Figure A5.3. Occupation probabilities for a two-level system with $\omega = \omega_0 = 8$ and $\tilde{\Omega} = 0.5$. The system starts in level 1, the lower energy level. $p_1(t)$ is solid and $p_2(t)$ is short-dashed for the full solution. $p_1(t)$ is dotted and $p_2(t)$ is long-dashed for the RWA solution.

16, 24 and 32, respectively. A close inspection of these figures reveals the full solution is more closely approaching the RWA solution with the increase in γ.

Reference

D. Bonacci, "Rabi spectra — a simple tool for analyzing the limitations of RWA in modelling of the selective population transfer in many-level quantum systems", arXiv: quant-ph/0309126v4, 7 May 2004.

Appendix 6. The Second-Order Correlation Function for a Two-Level System

Here, we present a derivation for $g^{(2)}(\tau)$ that is based on Loudon (2000, Secs. 4.9 and 7.9) and leads to Eq. (6.3.1). This approach is based on operators, the Interaction Picture and the Quantum Regression Theorem. We first introduce the operators and develop expressions for products of the time-dependent operators, which we relate to the time-dependent ρ_{ij}. These are then used to express the two-time version of the second-order coherence in terms of $\rho_{11}(\tau)$ and Eq. (6.3.1) follows. Alternative derivations are found in Carmichael (1999, Chap. 2) and Rand (2010, Chap. 6).

We start with the operators that take the two-level system from its ground state, energy level 0, to its excited state, energy level 1, and vice versa. These are in terms of bras and kets and we assume the kets $|0\rangle$ and $|1\rangle$ are orthonormal. First,

$$\hat{\Pi} = |0\rangle\langle 1|, \tag{A.6.1}$$

takes the excited state to the ground state and the Hermitian conjugate operator

$$\hat{\Pi}^{\dagger} = |1\rangle\langle 0|, \tag{A.6.2}$$

takes the ground state to the excited state. These operations are verified by:

$$\hat{\Pi}^{\dagger}|0\rangle = |1\rangle\langle 0|0\rangle = |1\rangle, \tag{A.6.3}$$

$$\hat{\Pi}^{\dagger}|1\rangle = |1\rangle\langle 0|1\rangle = 0, \tag{A.6.4}$$

$$\hat{\Pi}|0\rangle = |0\rangle\langle 1|0\rangle = 0, \tag{A.6.5}$$

$$\hat{\Pi}|1\rangle = |0\rangle\langle 1|1\rangle = |0\rangle. \tag{A.6.6}$$

The products of these operators yield

$$\hat{\Pi}^\dagger\hat{\Pi} = |1\rangle\langle 0|0\rangle\langle 1| = |1\rangle\langle 1|, \tag{A.6.7}$$

and

$$\hat{\Pi}\hat{\Pi}^\dagger = |0\rangle\langle 1|1\rangle\langle 0| = |0\rangle\langle 0|. \tag{A.6.8}$$

These combine to yield the identity operator $\hat{I}$. The operators in Eqs. (A.6.7) and (A.6.8) lead to the populations of the excited and ground states, respectively.

In the Interaction Picture the operators have the time dependences

$$\hat{\Pi}(t) = \hat{\Pi}e^{-i\omega_0 t}, \tag{A.6.9}$$

$$\hat{\Pi}^\dagger(t) = \hat{\Pi}^\dagger e^{i\omega_0 t}, \tag{A.6.10}$$

where ω_0 is the angular frequency, corresponding to the energy difference between the excited state and the ground state, when there is no electromagnetic field. We follow Loudon (2000, Sec. 7.9) and write

$$\langle\hat{\Pi}(t)\rangle = \rho_{10}(t)e^{-i\omega_0 t}, \tag{A.6.11}$$

$$\langle\hat{\Pi}^\dagger(t)\rangle = \rho_{01}(t)e^{i\omega_0 t}, \tag{A.6.12}$$

and

$$\langle\hat{\Pi}^\dagger(t)\hat{\Pi}(t)\rangle = \rho_{11}(t). \tag{A.6.13}$$

With these time dependences, we now turn to the two-time version of the second-order coherence in which we replace the electric field operators by the above operators

$$g^{(2)}(t,\tau) = \frac{\langle\hat{\Pi}^\dagger(t)\hat{\Pi}^\dagger(t+\tau)\hat{\Pi}(t+\tau)\hat{\Pi}(t)\rangle}{\langle\hat{\Pi}^\dagger(t)\hat{\Pi}(t)\rangle^2}, \tag{A.6.14}$$

and we are summing over the states of our two-level system. Our interest is in continuous-wave illumination, so we will eventually let $t \to \infty$ and determine $g^{(2)}(\tau)$.

With these preliminaries, we now need to express Eq. (A.6.14) in terms of the time-dependent ρ_{11}. We have four density matrix elements, but only three are independent due to

$$\rho_{00}(t) + \rho_{11}(t) = 1. \qquad (A.6.15)$$

We take the time dependence to be a sum and introduce the relation

$$\rho_{11}(t+\tau) = \beta_1(\tau) + \beta_2(\tau)\rho_{10}(t) + \beta_3(\tau)\rho_{01}(t) + \beta_4(\tau)\rho_{11}(t). \qquad (A.6.16)$$

The partial derivative of $\rho_{11}(t+\tau)$ with respect to t, equals the partial derivative of $\rho_{11}(t+\tau)$ with respect to τ, and this motivates the separable expression of Eq. (A.6.16). Further development follows the application of the Liouville–von Neumann Equations. We note Eq. (A.6.16) holds numerically for the case of Sec. 6.3, so we press forward. The $\beta_i(\tau)$ are seen to satisfy $\beta_1(0) = \beta_2(0) = \beta_3(0) = 0$, and $\beta_4(0) = 1$. Finally, we use Eqs. (A.6.11) to (A.6.13) to rewrite Eq. (A.6.16)

$$\langle \hat{\Pi}^\dagger(t+\tau)\hat{\Pi}(t+\tau)\rangle = \beta_1(\tau) + \beta_2(\tau)\langle\hat{\Pi}(t)\rangle e^{i\omega_0 t}$$
$$+ \beta_3(\tau)\langle\hat{\Pi}^\dagger(t)\rangle e^{-i\omega_0 t} + \beta_4(\tau)\langle\hat{\Pi}^\dagger(t)\hat{\Pi}(t)\rangle. \qquad (A.6.17)$$

The left-hand side of Eq. (A.6.17) has averages with two times, while the averages on the right-hand side involve one time. This brings in the Quantum Regression Theorem of Onsager (1931) and Lax (1963), which is discussed in Barnett and Radmore (1997, Chap. 5). This theorem lets us recast Eq. (A.6.17) as

$$\langle \hat{\Pi}^\dagger(t)\hat{\Pi}^\dagger(t+\tau)\hat{\Pi}(t+\tau)\hat{\Pi}(t)\rangle = \beta_1(\tau)\langle\hat{\Pi}^\dagger(t)\hat{\Pi}(t)\rangle$$
$$+ \beta_2(\tau)\langle\hat{\Pi}^\dagger(t)\hat{\Pi}(t)\hat{\Pi}(t)\rangle e^{i\omega_0 t} + \beta_3(\tau)\langle\hat{\Pi}^\dagger(t)\hat{\Pi}^\dagger(t)\hat{\Pi}(t)\rangle e^{-i\omega_0 t}$$
$$+ \beta_4(\tau)\langle\hat{\Pi}^\dagger(t)\hat{\Pi}^\dagger(t)\hat{\Pi}(t)\hat{\Pi}(t)\rangle, \qquad (A.6.18)$$

where we have introduced the additional operators $\hat{\Pi}^\dagger(t)$ and $\hat{\Pi}(t)$ in each term. We recognize the coefficient of $\beta_1(\tau)$ is $\rho_{11}(t)$ according to Eq. (A.6.13). Now, we simply need to evaluate the last three terms on the right-hand side of this equation!

The definitions of the operators allow us to realize the second term involves

$$\hat{\Pi}\hat{\Pi} = |0\rangle\langle 1|0\rangle\langle 1| = 0, \tag{A.6.19}$$

and the third term has

$$\hat{\Pi}^\dagger\hat{\Pi}^\dagger = |1\rangle\langle 0|1\rangle\langle 0| = 0. \tag{A.6.20}$$

These results also allow us to drop the fourth term. We are left with

$$\langle\hat{\Pi}^\dagger(t)\hat{\Pi}^\dagger(t+\tau)\hat{\Pi}(t+\tau)\hat{\Pi}(t)\rangle = \beta_1(\tau)\langle\hat{\Pi}^\dagger(t)\hat{\Pi}(t)\rangle. \tag{A.6.21}$$

We set $t = 0$ in Eq. (A.6.16) and use our assumed initial conditions

$$\rho_{00}(t = 0) = 1, \tag{A.6.22}$$

$$\rho_{01}(t = 0) = \rho_{10}(t = 0) = \rho_{11}(t = 0) = 0, \tag{A.6.23}$$

to find

$$\rho_{11}(\tau) = \beta_1(\tau). \tag{A.6.24}$$

With these simplifications, Eq. (A.6.14) becomes

$$g^{(2)}(t,\tau) = \frac{\rho_{11}(\tau)\langle\hat{\Pi}^\dagger(t)\hat{\Pi}(t)\rangle}{\langle\hat{\Pi}^\dagger(t)\hat{\Pi}(t)\rangle^2} = \frac{\rho_{11}(\tau)}{\langle\hat{\Pi}^\dagger(t)\hat{\Pi}(t)\rangle} = \frac{\rho_{11}(\tau)}{\rho_{11}(t)}. \tag{A.6.25}$$

The final step is to let $t \to \infty$, so

$$g^{(2)}(\tau) = \frac{\rho_{11}(\tau)}{\rho_{11}(\infty)}, \tag{A.6.26}$$

and we have the form of the second-order correlation function for a two-level system we assumed in Eq. (6.3.1).

References

S. M. Barnett and P. M. Radmore, *Methods in Theoretical Quantum Optics* (Oxford University Press, Oxford, 1997).

H. J. Carmichael, *Statistical Methods in Quantum Optics 1* (Springer-Verlag, Berlin, 1999).

M. Lax, "Formal theory of quantum fluctuations from a driven state", *Phys. Rev.* **129**, 2342–2348 (1963).

R. Loudon, *The Quantum Theory of Light*, 3rd ed. (Oxford University Press, Oxford, 2000).

L. Onsager, "Reciprocal relations in irreversible processes I", *Phys. Rev.* **37**, 405–429 (1931).

S. C. Rand, *Lectures on Light: Nonlinear and Quantum Optics using the Density Matrix* (Oxford University Press, Oxford, 2010).

Index